AF443412

An introduction to
Environmental Pattern Analysis

Computer programs to carry out the main methods discussed in this book can be found in the chapter by P.J.A. Howard in the book *Microcomputers in Environmental Biology* edited by Professor J.N.R. Jeffers and published by Parthenon Publishing Ltd.

An introduction to
Environmental Pattern Analysis

P.J.A. Howard

The Parthenon Publishing Group

International Publishers in Medicine, Science & Technology

Casterton Hall, Carnforth,
Lancs, LA6 2LA, UK

120 Mill Road, Park Ridge,
New Jersey 07656, USA

Published in the UK by
The Parthenon Publishing Group Limited
Casterton Hall, Carnforth
Lancs, LA6 2LA, England

Published in the USA by
The Parthenon Publishing Group Inc.
120 Mill Road,
Park Ridge,
New Jersey 07656, USA

British Library Cataloguing in Publication Data
Howard, P. J. A.
 An introduction to environmental pattern analysis
 1. Patterns. Mathematical aspects
 I. Title
 516

 ISBN 1-85070-240-3

Library of Congress Cataloging-in-Publication Data
Howard, P. J. A. (Peter John Arthur)
 An introduction to environmental pattern analysis / by P. J. A. Howard
     p.   cm.
 Includes bibliographical references and index.
 ISBN 1-85070-240-3 : $85.00
 1. Pollution–Measurement.   2. Cluster analysis.   I. Title.
 TD193.H69   1990
 363.7'02–dc20   90-45034
                                                                      CIP

First published 1991

Phototypesetting by Lasertext Ltd., Manchester
Printed and bound in Great Britain by
Cromwell Press Ltd., Broughton Gifford, Wiltshire

CONTENTS

Section 1

Introduction and basic concepts

CHAPTER 1

INTRODUCTION

Increasing concern about the effects of mounting pressures on the environment, together with changes in land use resulting from evolving agricultural policies, are resulting in increased use of computers to analyse environmental data. In particular, a researcher is often faced with the problem of interpreting a multivariate set of observations. Commonly, especially with survey data, the problem is not statistical in the sense of hypothesis testing, rather it is the question of reducing a complex set of data to an efficient summary, a process which may result in the generation of hypotheses. If we have observations on N objects over a single battery of p attributes, the problem may be one of searching for structure in the data. This may involve investigating how the attributes measured for a sample are interrelated and whether they occur in different combinations in objects representing different aspects of the population.

Alternatively, we may wish to know how the objects themselves are interrelated, perhaps with the idea of searching for groupings which may be useful either practically or in interpretation. For this type of study, one of the ordination techniques might be appropriate, possibly followed by some form of cluster analysis. If there is an *a priori* division of the attributes into two batteries, then, provided that the theoretical requirements are met, we can use the technique of canonical correlation, which searches for pairs of linear functions that maximally correlate but which themselves are uncorrelated with previously-calculated pairs.

If there is an *a priori* division of the objects into groups, we can study the relationships between the groups by discriminant analysis or by canonical variate analysis. In recent years, such applications have come to be known informally as pattern analysis.

Many of the techniques have been widely used in the taxonomy of organisms, and terms and concepts from that field will be introduced where appropriate. The development of numerical methods in taxonomy had several effects. What used to be an intuitive art became formalized into a quantitative science. The increasing availability of digital computers meant that it became practicable to explore the use of a wide range of numerical techniques, and this attracted the attention of statisticians and mathematicians, with the consequent development of a wide variety of methods and their application to a variety of problems in different fields. This has not been an unmixed blessing, as there now exists a bewildering variety of numerical techniques, the properties of many being not fully known. Numerical methods can help the researcher to investigate the structure of his/her data, but the results of the analyses still need to be interpreted. It is, perhaps, worth bearing in mind the three questions: Why do it? How do you do it? When you have done it, what does it mean?

It is easy to perceive structure in data when the structure and discontinuities are obvious, but such a situation is not typical. Much of what we observe in nature changes continuously in one property or another, but not necessarily with equally steep gradients for each property. It is such cases which give the researcher the greatest problems in deciding where or how to draw boundaries, or even whether boundaries should be drawn at all. One aim of this book is to clarify some of the issues involved in the use of numerical methods to search for structure in environmental data, and, in particular, in those methods used for classification.

The use of classification methods with environmental data is much less well-developed than is the taxonomy of organisms. This is due partly to the diversity of interests of environmental researchers and partly to the nature of environmental data, which do not often lend themselves to easy classification. The appropriate methods in the taxonomy of organisms are not necessarily the most appropriate for environmental problems (Clifford and Stephenson 1975). A theoretical approach is taken because it is necessary to understand the theory of the methods in order to understand how they should be used and what their limitations are. Indeed, the same is true for multivariate methods in general. By virtue of the speed with which they can process large quantities of data and perform complex calculations, computers have revolutionized the handling of scientific data. However, there is the danger that the power of computers sometimes leads those unfamiliar with the theory of the methods to assume that a computer output contains some kind of objective 'truth' which should not be questioned.

Thus, with regard to cluster analysis, Anderberg (1973) commented: 'Cluster analysis methods involve a mixture of imposing a structure on the data and revealing that structure which actually exists in the data. The notion of finding natural groups tends to imply that the algorithm should passively conform like a wet tee shirt. Unfortunately, practical procedures involve fixed sequences of operations which systematically ignore some aspects of the structure while intensively dwelling on others'. The use of a computer does not release the ecologist from the necessity of thinking, quite the reverse. A computer output will present the ecologist with the necessity to decide what the computer's results mean in terms of the data being analysed and the problem to be solved. Multivariate methods lay traps for the unwary; it is a second aim of this book to spring many of them.

Essentially, what we are discussing is exploratory data analysis, the examination of the data from various angles and the piecing together of the information obtained in order to generate hypotheses which can be tested by more formal methods. As Mather (1976) pointed out, learning comes from the interplay between the investigator, the technique, and the data. Such an interplay is unlikely to be fruitful unless the investigator has a good basic knowledge of the numerical techniques as well as a firm grasp of the ecological system under consideration. A particular feature of this book is Chapter 16, on choosing a strategy for data analysis. That chapter will enable a researcher to choose an appropriate method or methods, and then to go to the appropriate chapter for more details.

The problem for many researchers is how to acquire a good basic knowledge of numerical techniques. Many books on multivariate analysis are highly theoretical, difficult for non-mathematical researchers to read, and do not contain much information which is of practical use in data analysis. Two books which can be recommended are Chatfield and Collins (1980) and Pimentel (1979). The latter is exceptionally good, and contains much practically useful information. Many papers published in journals which are unlikely to be read by many ecologists contain information on practical applications which are not usually given in standard textbook treatments. An example of this is the use of typical points in principal component analysis. A third aim of this book is to draw the attention of the reader to such sources of information, and to give sufficient of the theoretical background to enable an ecologist to acquire the good basic knowledge required.

CHAPTER 2

SOME BASIC MATHEMATICAL AND
NUMERICAL CONCEPTS

To enable non-mathematical researchers to follow the theoretical discussions of the methods, it is useful to define some terms and to explain some relevant algebra and geometry. We shall not consider in detail the basic algebraic manipulation of vectors and matrices as there are many excellent textbooks to which the reader can refer (e.g. Searle 1966). Instead, we shall concentrate on those aspects which are not always made explicit but which are important to an understanding of the methods to be discussed. For a more detailed discussion, see Jöreskog *et al.* 1976.

2.1 Definitions and notation

Objects and attributes

We shall use the term *objects* to represent the things upon which the observations have been made. The objects may be diverse, for example they may be samples of soil on which chemical analyses have been made, areas of land varying in size from a square metre quadrat to a 10 km × 10 km map grid square on which observations have been made, or individual organisms.

A *variable* is a symbol used to represent some specified set of values. Any member of that set is a value of the variable, and the set itself is the range of the variable. The term may be used to denote non-measurable characteristics, e.g. male or female. An *attribute* is a qualitative characteristic the presence or absence of which may assign a quantitative value to a variable (James and James 1968; Kendall and Buckland 1971). Thus, the presence of species may be scored 1 and its absence 0. Qualitative characteristics may be fundamentally quantitative, e.g. a species which is absent, rare, common, or abundant, may be scored qualitatively if it is not possible to measure its abundance. Hence, there is no clear distinction between the terms 'variable' and 'attribute'. For convenience, the term 'attribute' will be used here, because the term 'variable' is not generally associated with qualitative characteristics. The different types of attribute will be discussed more fully in the next chapter.

7

A *variate* is a quantity which may take any of the values of a specified set with a specified relative frequency or probability. It is often known as a random variable.

Matrix

A matrix is an array of numbers with one or more rows and one or more columns. An individual entry in the matrix is called an *element*. Each element is addressed by the number of its row (i) and its column (j). The order of the matrix is the number of rows and the number of columns, and is often given after the letter symbolizing the matrix, thus:

$$\mathbf{X}\ (4 \times 3) = \begin{bmatrix} x_{11} & x_{12} & x_{13} \\ x_{21} & x_{22} & x_{23} \\ x_{31} & x_{32} & x_{33} \\ x_{41} & x_{42} & x_{43} \end{bmatrix}$$

The *transpose* of a matrix is the original matrix with all the rows and columns interchanged. The transpose is symbolized by attaching a prime to the matrix symbol, thus:

$$\mathbf{X}\ (4 \times 2) = \begin{bmatrix} 7 & 5 \\ 9 & 8 \\ 4 & 3 \\ 6 & 7 \end{bmatrix} \qquad \mathbf{X}'\ (2 \times 4) = \begin{bmatrix} 7 & 9 & 4 & 6 \\ 5 & 8 & 3 & 7 \end{bmatrix}$$

A matrix may be *rectangular* or *square*. For a square matrix, in which the number of rows is equal to the number of columns, the order is usually given simply as a single figure. In a square matrix $\mathbf{A}$, the elements a_{11}, a_{22}, a_{33} and so on, are called the *principal diagonal* elements, a_{11} being the *leading term*. The sum of these diagonal elements is called the *trace* of the matrix. A *diagonal* matrix is a square matrix with all off-diagonal (i.e. off principal diagonal) elements zero. An *identity* matrix, usually given the symbol $\mathbf{I}$, is a diagonal matrix in which elements in the principal diagonal are all unity. A square matrix is *symmetric* if $a_{ij} = a_{ji}$ for all $i, j = 1$ to p, where p is the order. Square symmetric matrices are important in many multivariate methods, a correlation matrix is an example.

Orthogonal matrix

A square, non-singular matrix $\mathbf{A}$ is said to be orthogonal if its inverse equals its transpose and

$$\mathbf{A}'\mathbf{A} = \mathbf{A}\mathbf{A}' = \mathbf{I}$$

Vector

A vector may be thought of as a single row or column. To symbolize a column vector, we use a heavy type lower-case letter; to symbolize the same vector as a row vector, we add a prime, thus:

$$\mathbf{x} = \begin{bmatrix} x_1 \\ x_2 \\ x_3 \end{bmatrix} \qquad \mathbf{x}' = [x_1\, x_2\, x_3]$$

Hence, a matrix may consist of several column and row vectors, each of which may be identified by the row or column subscript.

Scalar

A scalar is a simple number, as distinct from a matrix or vector.

Linear independence of vectors and rank of a matrix

A vector that is a scalar multiple of another vector or is a sum (weighted or unweighted) of a set of vectors, is said to be *linearly dependent*. Conversely, a set of vectors is said to be *linearly independent* if no vector is a scalar multiple of any other vector, or a combination of any set of the other vectors.

Suppose that a matrix $\mathbf{X}$ ($N \times p$) is a product of two other matrices:

$$\mathbf{X}(N \times p) = \mathbf{A}(N \times r) \times \mathbf{B}(r \times p)$$

The *rank* of $\mathbf{X}$ is the smallest common order among all pairs of matrices whose product is $\mathbf{X}$. If r is the smallest common order, it is the rank of $\mathbf{X}$. Clearly, it is not possible for the rank of a matrix to exceed its smallest dimension, but it can be less. The rank of a matrix will be equal to its smallest dimension if the vectors numbered in that dimension are linearly independent. In the above example, if $N > p$ and $r = p$ then the column vectors of $\mathbf{X}$ are linearly independent and the matrix is said to be of full rank. If $r < p$, then r is the number of linearly independent vectors. The rank of a data matrix is an important property.

Major and minor product moments and the rank of a matrix

When a matrix is post-multiplied by its own transpose, the result is a major product moment. Thus, for matrix $\mathbf{X}$ ($N \times p$), the major product moment matrix $\mathbf{A}$ ($N \times N$) is symmetric:

$$\mathbf{A} = \mathbf{X}\mathbf{X}'$$

A minor product moment is obtained by pre-multiplying a matrix by its transpose. Thus, for a data matrix $\mathbf{X}$ ($N \times p$) in which each element is expressed as a deviation from the column mean, the minor product moment matrix $\mathbf{W}$ ($p \times p$), given by

$$\mathbf{W} = \mathbf{X}'\mathbf{X}$$

is symmetric and is a sum of squares and products matrix. A correlation matrix is the minor product moment of the standardized data matrix with each element divided by $N - 1$.

The rank of a product moment matrix is equal to the rank of its factor. Since the number of non-zero eigenvalues of a square matrix is equal to its rank, a square matrix of order p and rank r will have r non-zero eigenvalues, with corresponding linearly independent vectors, and $(p - r)$ zero eigenvalues with corresponding linearly dependent vectors.

Norms of vectors and matrices

Sometimes it can be useful to have a scalar measure of the magnitude of a vector. Such a measure is called a *norm*, and for a vector $\mathbf{a}$ the norm is written as $\|\mathbf{a}\|$. Normalization of a real vector produces a unique result, and so can be used in comparison of results. A useful measure is the *Euclidean norm*:

$$\|\mathbf{a}\|_E = \left[\sum_{i=1}^{p} a_i^2 \right]^{1/2} = (\mathbf{a}'\mathbf{a})^{1/2}$$

Thus, if a vector $\mathbf{a}$ is pre-multiplied by its transpose, the resulting inner product (or minor product moment) $\mathbf{a}'\mathbf{a}$ represents the sum of the squared vector elements. As we shall see below, this is a widely-used basis of normalization of vectors. When the Euclidean norm equals unity, the vector has length 1 and is called a unit vector. The important property of a vector is the relationship between the vector elements. If a vector is multiplied, or divided, by a scalar this relationship is unaltered, but the length of the vector is changed. In ordination methods it is frequently convenient to change the length of vectors. A unit vector is obtained by dividing each element of the vector by the Euclidean norm.

The Euclidean norm for a matrix $\mathbf{A}$ ($p \times q$) is:

$$\|\mathbf{A}\|_E = \left[\sum_{i=1}^{p} \sum_{j=1}^{q} a_{ij}^2 \right]^{1/2}$$

that is, the square root of the sum of the squared elements of the matrix.

2.2 Geometry of vectors

We have defined a vector algebraically, but a vector also has a geometrical interpretation. It may be defined as a directed line segment. Let us consider a data matrix $\mathbf{X}$ ($N \times p$). We shall use the convention that the rows of a

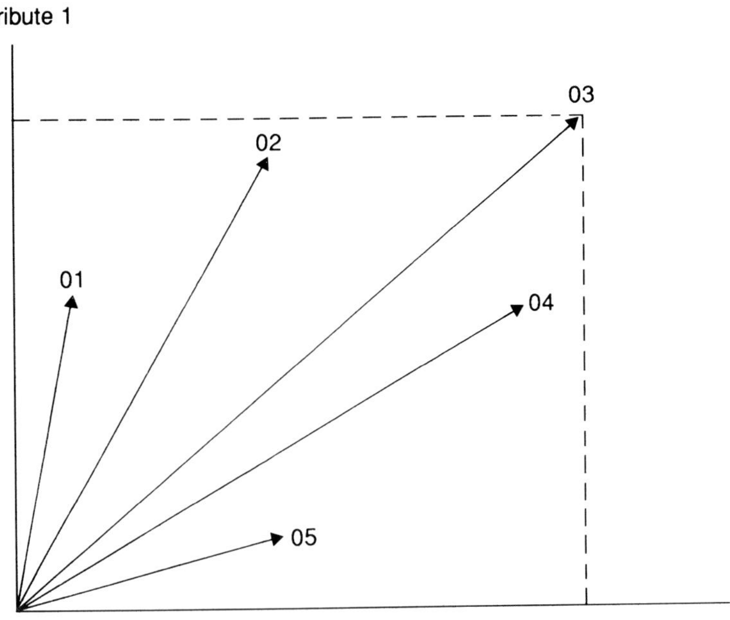

Figure 2.1 Representation of objects as vectors in a two-dimensional attribute space

data matrix represent the N objects, and the columns represent the p attributes. If we use the attributes as co-ordinate axes and plot the positions of the objects as in Figure 2.1, the objects are said to occupy the space defined by the attributes, i.e. the attribute space. The line joining each point to the origin is the geometrical representation of the corresponding vector, i.e. a row vector of the data matrix. The N objects occupy the p-dimensional attribute space. Similarly, we could plot the positions of the attributes using the objects as co-ordinate axes, as in Figure 2.2. In this case, the line joining each point to the origin is the geometrical representation of a column vector in the data matrix, and the p attributes occupy the N-dimensional object space.

Since vectors can be represented by lines, it is clear that the angles between the lines can be found. The cosine of the angle between two vectors $\mathbf{v}$ and $\mathbf{w}$ is the inner product of the unit vectors:

$$\cos \theta = \frac{\mathbf{v'w}}{(\mathbf{v'v})^{1/2}(\mathbf{w'w})^{1/2}}$$

Two unit vectors are orthogonal if their inner product is zero, that is, if $\cos \theta = 0$, so that $\theta = 90°$. Vectors which are both orthogonal and normalized to unit length are said to be orthonormal.

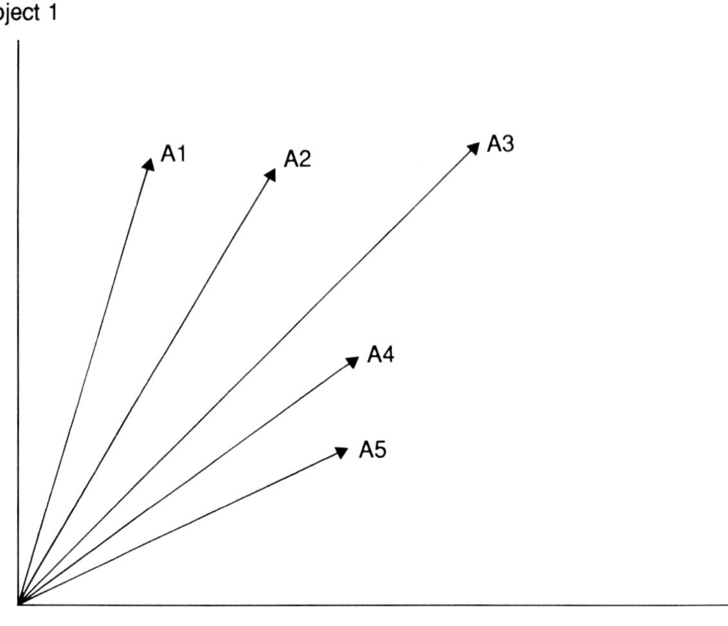

Figure 2.2   Representation of attributes as vectors in a two-dimensional object space

Linearly dependent and independent vectors

If a vector is linearly dependent on a set of other vectors, it will lie in the same plane as those vectors. The dimensionality of the subspace in which the vectors lie is the same as the number of linearly independent vectors in the set. The latter are said to form a basis for this subspace. However, it should be noted that they are not the only basis since, in any plane in which the vectors lie, we could consider any pair of orthogonal vectors to be a basis, and such vectors could be rotated to an infinity of positions in the same plane and still form a basis for the subspace.

Therefore, if we have a matrix $\mathbf{X}$ ($N \times p$) of rank r, where $r < p$, then only r of the p column vectors are linearly independent and these r vectors will form a basis of an r-dimensional subspace (of the original N-space) in which the p column vectors will lie. Any set of r orthogonal vectors could provide a basis for the subspace. This concept is important in factor analysis.

Variance and correlation

Let us consider a matrix $\mathbf{X}$ ($N \times p$) of data expressed as deviations from the variate means. The variance of the jth variate is given by:

$$s_j^2 = \frac{\sum\limits_{i=1}^{N} x_{ij}^2}{N-1} = \frac{\mathbf{x}_j' \mathbf{x}_j}{N-1} = \frac{\|\mathbf{x}_j\|^2}{N-1}$$

Hence, the standard deviation, which is given by

$$s_j = \frac{\|\mathbf{x}_j\|}{\sqrt{N-1}}$$

is proportional to the length of the vector of observations on the variate expressed as deviations from the variate mean.

Standardization of $\mathbf{X}$, to give each variate a mean of zero and a standard deviation of unity, gives matrix $\mathbf{Z}$, i.e.

$$z_{ij} = x_{ij}/s_j$$

Standardized variates $\mathbf{z}_j$ are represented by vectors of length $\sqrt{(N-1)}$. The angle between two standardized variates is given by:

$$\cos \theta = \frac{\mathbf{z}_j}{\sqrt{N-1}} \times \frac{\mathbf{z}_k}{\sqrt{N-1}} = \frac{\mathbf{z}_j' \mathbf{z}_k}{N-1} = \frac{\sum\limits_{i=1}^{N} z_{ij} z_{ik}}{N-1}$$

which is the Pearson product-moment correlation coefficient.

2.3 Q-mode and R-mode analysis

The difference between Q-mode and R-mode analyses is important in pattern analysis. Referring to our data matrix $\mathbf{X}$ ($N \times p$), then, basically, R-mode analysis is the analysis of the minor product moment $\mathbf{X}'\mathbf{X}$. As we have already seen, a product-moment correlation matrix is a minor product moment matrix, so that an analysis based on a correlation matrix is an R-mode analysis. Clearly, such an analysis is based on the relationships among the variables, i.e. among the N rows of the data matrix over the p columns. Conversely, Q-mode analysis is the analysis of the major product moment $\mathbf{X}\mathbf{X}'$. Here we are comparing the numbers of individuals shared by two attributes, i.e. we are examining the relationships among the p columns over the N rows of the data matrix. A similarity matrix (see Chapter 12) is a Q-mode matrix.

In vegetation analysis, the classification of sites or stands using species as attributes has been called a *normal* analysis, while the classification of species in terms of the sites in which they occur has been called an *inverse* analysis (Lambert and Dale 1964). Williams and Dale (1965) suggested that while the terms R and Q refer to the type of matrix being analysed, some terms are necessary to describe the underlying model, i.e. the space in which the user is interested. They suggested the terms A (for the model in the attribute-space) and I (for the model in the individual- or object-space). This also introduces the problem of centring and scaling of the data (Orloci 1967b; Noy-Meir 1973). In vegetation analysis, R and Q were subsequently treated as indicating that either an interspecies or interstand coefficient is used with species-centred data or without regard to the underlying geometrical representation (Dale 1975; see also Orloci 1967b, and Ivimey-Cook *et al.* 1969). Such conceptual spaces were also discussed by Gauch and Whittaker (1981).

Singular value decomposition, or basic structure, of a rectangular matrix

The singular value decomposition (Eckart and Young 1936; Johnson 1963) states that a rectangular matrix $\mathbf{X}$ ($N \times p$), where $N > p$ and rank $r < p$, can be represented as

$$\mathbf{X} = \mathbf{WGA}' \tag{2.1}$$

where $\mathbf{W}$ ($N \times r$) has orthonormal columns, $\mathbf{A}$ ($p \times r$) has orthonormal columns, and $\mathbf{G}$ ($r \times r$) is a diagonal matrix, the diagonal elements of which are positive and are called the singular values of $\mathbf{X}$. Equation (2.1) can be re-written as

$$\mathbf{X} = g_1\mathbf{w}_1\mathbf{a}_1' + g_2\mathbf{w}_2\mathbf{a}_2'\ldots\ldots + g_r\mathbf{w}_r\mathbf{a}_r' \tag{2.2}$$

that is, matrix $\mathbf{X}$, of rank r, is a linear combination of r matrices of unit rank.

The minor product moment $\mathbf{X}'\mathbf{X}$ and major product moment $\mathbf{XX}'$ have the same positive eigenvalues, and the singular values are the positive square roots of those eigenvalues. The eigenvectors of $\mathbf{X}'\mathbf{X}$ are the columns of $\mathbf{A}$, and the eigenvectors of $\mathbf{XX}'$ are the columns of $\mathbf{W}$. The singular value decomposition shows how the results of a Q-mode analysis are related to those of an R-mode analysis, from (2.1):

$$\mathbf{W} = \mathbf{XAG}^{-1} \tag{2.3}$$

$$\mathbf{A} = \mathbf{X}'\mathbf{WG}^{-1} \tag{2.4}$$

In practical computation, the great virtue of the singular value decomposition is that it can be performed in such a stable manner that rounding errors do not bring any significant complications (Wilkinson 1978). The singular values of a matrix are very stable with respect to changes in the elements of the matrix (Lawson and Hanson 1974). Because the singular value decomposition can operate on rectangular matrices, which need not be square or symmetric, it can be useful in various multivariate methods, e.g. correspondence analysis.

For computers with limited storage, or lacking double-precision arithmetic, singular value decomposition can prove the most, if not the only, efficient method of obtaining a multiple regression or a principal component analysis if large numbers of variates are involved (Banfield 1978). If matrix **W** is not needed, one might simply solve the eigenstructure of **X'X**. However, the explicit computation of **X'X** involves unnecessary numerical inaccuracy (Golub and Reinsch 1971).

2.4 Errors in computation

Computer users, especially those writing their own programs, should always bear in mind the fact that errors in computing can and do occur. In addition to possible inaccuracies in the preparation of data, we can recognize two main ways in which errors can occur in the manipulation of the data by the computer: (i) errors of machine representation of floating point numbers, and (ii) errors in arithmetical operations. The extent and importance of these two types of error depend on the length and structure of the computer word and the manner in which the arithmetical operations are executed.

The way that we do arithmetic is closely related to the way in which we represent numbers. In positional notation, a number has a base, or radix, b:

$$ABCD.EFG = Ab^3 + Bb^2 + Cb + D + Eb^{-1} + Fb^{-2} + Gb^{-3}$$

The dot is called the radix point. Mostly, scientists use the decimal system with $b = 10$. Computers may use base 2, 4, 8, 10 or 16. In handling such numbers in a computer, it is convenient to let the position of the radix point 'float' as a program is running, and to carry with each number an indication of the radix point position. In a computer, a floating-point number is represented by a sign, a fraction (mantissa), and an exponent (actually a characteristic representing the exponent) which indicates the radix point position. The word length of the computer limits the number of digits in the fraction. In the DEC RSTS/E system, for example, the base is 2, i.e. the numbers are stored in binary form. In single-precision format, this system stores floating-point numbers in two words, each of 16 binary digits (bits), which can accurately represent numbers with up to 6 decimal digits. Numbers with more than 6 digits are rounded in this system, although in some systems they are chopped.

In decimal representation, a significant figure is any one of the digits 1 to 9, and 0 is a significant figure unless it is used to fix the position of the decimal point. The number 0.01517 could equally be represented as 0.1517×10^{-1}. Clearly, if we have only a limited number of decimal digits stored, we represent the number more accurately by making the most significant digit, i.e. the one nearest the decimal point, non-zero. In computing terminology, such a floating-point number is said to be normalized.

The absolute error of an approximate number Y is $Y - X$, where X is the true value we are trying to represent. The relative error is $(Y - X)/X$. Sometimes we use an approximate number because we cannot write the exact number as a decimal, e.g. π, 2/3, and some other fractions, or because we cannot measure it to the ultimate degree of accuracy. A significant digit in an approximate number can be referred to as 'correct' if rounding the approximate number to just after that digit position will cause the absolute error to be no more than half a unit in that digit position. The relative error of an approximate number is closely related to the number of correct significant digits.

Floating-point computation is by nature inexact, and it is not difficult to misuse it so that the computed answers consist almost entirely of 'noise'. Roughly speaking, the operations of floating-point multiplication and division do not magnify the relative error by very much. On the other hand, floating-point subtraction of nearly equal quantities, and floating-point addition of quantities which have nearly equal absolute values but opposite sign, can increase very greatly the relative error. This is the greatest source of inaccuracy in most calculations, and it forms the weakest link in a chain computation where it occurs. If the calculation is iterative, the result can be disastrous. In statistics, this type of inaccuracy must be borne in mind when calculating sums and sums of products, for example in the calculation of covariance and correlation matrices. Algorithms especially designed to overcome this problem are required (e.g. Youngs and Cramer 1971; Malcolm 1971).

More detailed, but very readable, accounts of these problems are given by Pennington (1970) and Henrich (1980), and a more technical account is given by Knuth (1969).

CHAPTER 3

THE NATURE OF THE DATA

3.1 Types of data

Data consist of attribute scores. Conventionally, in statistics, the term 'attribute' is usually used for qualities possessed, or not possessed, by an object. The term 'variable' is usually used for quantities which vary continuously. However, in pattern analysis the term 'attribute' has come to be used in a wider sense. This is convenient, since we do not need to differentiate between continuous and discrete data in general discussions where the nature of the data is not in question. There are many different kinds of attributes, see for example Sneath and Sokal 1973; Clifford and Stephenson 1975; Williams 1976. The most common kinds of attributes are: (i) binary, e.g. presence–absence; (ii) disordered multistate (also called nominal attributes), such as colour or rock type; (iii) ordered multistate (also called ordinal attributes, e.g. rare, common, abundant; (iv) meristic, e.g. number of petals; (v) continuous, i.e. measures on a continuous scale (also called quantitative or numeric attributes).

The scale on which the attributes are scored or measured will have an important influence on the subsequent data analysis. On a *nominal* scale, observations are recorded in mutually-exclusive categories of equal rank. For presence–absence data an object can be in either state, usually scored 0 or 1. With disordered multistate data, a given object can be referred to only one state. The states may be numbered for computational convenience, but no meaning can be attached to the order in which the states are taken. An *ordinal* scale is used when various levels can be established for an attribute, but the scale values establish only the order of the observations, they do not contain any information on relative distances. With ordered multistate data, for example, rare, common, abundant, could be coded as 1, 2, 3, but these scores would not represent the relative abundancies, i.e. the distances between the states are undefined. With *interval* and *ratio* scales, there is a concept of distance. On an interval scale, both the order

17

and the magnitude of an attribute state can be found relative to some arbitrary zero value, as in temperature scales. A ratio scale is used when the order and magnitude of an attribute state can be referred to some natural origin with no negative values, as in measurement of length or weight. On such a scale, the ratios between scale values are meaningful, as are the sums and differences. Probably few biologists have much formal background in types of measurement and scales. Torgerson's (1958) book is useful for this.

Numbers carry the least information when they serve only as labels. Hence, the kinds of statistical operations that can logically be performed on disordered multistate data are rather limited. As a simple example, if we coded male as 1 and female as 2, then, with a mixture of males and females, the mean code would be between 1 and 2, which is biologically meaningless. The only useful calculations which can be performed on such data are those based on counting, i.e. frequencies, fractions, and the percentages of objects in a coded category. The only permissible statistical summary is the mode, which is determined solely on the basis of the category with the greatest frequency.

In terms of information content, the next step up is to data on an ordinal scale, of which Moh's scale of mineral hardness and the Beaufort wind scale are examples. Because of the greater information content, data on an ordinal scale (ordered multistate data) are susceptible to a greater variety of statistical operations than are data on a nominal scale. In addition to frequencies, percentages, and modes, it is possible to determine whether any given data item is greater than, or less than, another, so that any operation based on ordering can be used. Percentiles can be used to compare relative frequencies of data items lying in major sections of the scale, and statistical summaries such as the median and ranges can be used. However, interpretation of small differences between medians and ranges of data sets is complicated by unequal scale intervals. The nature of ordinal scales makes calculation of means and standard deviations inappropriate, since the basis of the computations assumes regular scale graduations which are unlikely to be found on an ordinal scale.

The greatest variety of statistical operations is possible with data measured on interval and ratio scales, but the coefficient of variation should not be used with data measured on interval scales. The coefficient of variation is independent of units of measurement, but not when the zero point changes between scales. Similarly, if scale intervals are uniform and there is no shift of zero between scales, a given pair of readings will stand in the same ratio to each other no matter what scale is used. This is not true when the zero point is moved.

In pattern analysis, as in ordinary statistics, there is a greater choice of algorithms for dealing with continuous attributes than with other types. Ordered and disordered multistate data can be a nuisance to handle, for many purposes there is really no entirely satisfactory way of treating such data numerically. Meristic attributes, which can take only integral values, can also be a nuisance. In many cases, they cannot sensibly be treated as continuous; consider a case of counts of floral parts in a sample which contains some plants with two petals and some with four. As Williams (1976) pointed out, finding that the mean is three implies a transfer from dicotyledons to monocotyledons. In such a case, the attribute could be coded as disordered multistate. On the other hand, in some cases the mean value is interpretable and the data could be treated as continuous for practical purposes.

Continuous attributes are rarely continuous in the strict sense, since measurements are always made with limited accuracy, and there is always some degree of rounding off. In practice, the difference between continuous and discrete attributes depends on the chance that different observations take the same value. For practical purposes, counts that follow a Poisson distribution with a large mean can be regarded as continuous, since only a small proportion of the observations will take any one value. On the other hand, a continuous attribute grouped so coarsely that only a few values actually occur must be treated differently, at least for some types of analysis. If an attribute takes a wide range of values, but has a concentration at one value, usually zero (e.g. counts of parasites on a host), it may be better to score it as ordered multistate, e.g. zero, low, medium, high (Marriott 1974).

Data may be coarsely grouped, either because measurements have been made with very limited accuracy, or are replaced by rough assessments such as low, medium, high. For many purposes, such grouping is not important. The assessment can be replaced by suitable scores, giving whatever weight is considered appropriate to the differences between the groups. In many of the classical multivariate methods, the central limit theorem then justifies treating them as if they were jointly normally distributed. However, in some applications care is needed. This is especially true in cluster analysis. If the aim is to find a useful or meaningful grouping of the data, a coarsely-grouped attribute may exert a disproportionate influence on the result (Marriott 1974).

One way of dealing with disordered multistate data is to replace them by dummy binary attributes. Thus, the colours red, white and blue could be coded as two binary attributes, one taking the value 1 for red and 0 for blue and white, the other taking the value 1 for white and 0 for red and blue. However, this method becomes rather cumbersome if there are many disordered multistate attributes, or many disordered states. This method can also be misleading if it is used in conjunction with an analysis that does not take into account the fact that the resulting binary attributes are correlated (e.g. some forms of cluster analysis). With many observations of this type it is preferable to base a cluster analysis on some sort of similarity or dissimilarity measure (Marriott 1974). However, the comments of Rubin (1967) given below should be taken into account.

Just as in univariate applications the standard error of a proportion can be used for significance tests and confidence intervals as if the proportion were normally distributed, so in the multivariate case the central limit theorem often justifies treating binary data as approximately multivariate normal. However, if all the data are of this type, other models are available, and some special methods have been evolved (Marriott 1974). Methods based on the analysis of contingency tables (e.g. correspondence analysis) are particularly suitable.

Differences of opinion exist on the value of binary data in ecological work, but the consensus of opinion seems to be that other data types are preferable (Clifford and Stephenson 1975 p39). In a botanical context, Greig-Smith (1964 p160) stated: 'We are, in fact, dealing with a population of individuals (if stands may be so regarded) which differ from one another in terms of continuous variables of which presence and absence are only a crude expression'. In most branches of ecology, the tendency is to regard dominance (by some measure) as important (see e.g. the papers in Section III of Ord *et al.* 1979). Smartt *et al.* (1974) compared the effects of recording vegetation data in different ways (presence–absence, frequency, percentage cover, biomass) on the outcome of a number of numerical classification procedures. The results showed that the main discriminating feature between the different data types was the extent to which differences in floristic composition were masked by differences in the overall bulk of the plant material. An important distinction was made between (a) unbounded data, where the measures are free to vary without constraint across both the quadrats and species; (b) partially bounded data, where the measure is bounded for each species in a quadrat but each quadrat total is unbounded; and (c) bounded data, where the quadrat total itself is constrained to a maximum value and the species values can vary only within that constraint. Quantitative measures which are sufficiently constrained to allow quantitative differences to operate primarily within a framework of floristic diversity were thought to be the most generally useful for ecological work concerned with plant–habitat relationships. On the other hand, partially-bounded measures which are species-constrained (e.g. frequency) may prove to be too similar, in their performance, to presence–absence data to justify the extra time and effort in collecting them.

Results of analyses using data with numerical values are often more informative than those using binary data. For example, Williams *et al.* (1973) found that while plant species presence–absence was adequate in a simple study involving only eight sites, for ten sites there was 'some advantage' in using numbers, while for 80 sites, quantitative data were distinctly preferable. Barkham (1968) found quantitative data to be more informative than presence–absence data in a study of the vegetation of

Cotswold beechwoods. Presence–absence data appear to work well when there are major differences in species distribution between sites, but they are not very useful for detailed studies of pattern if there are relatively few species with less clear-cut differences between sites (cf. Smartt *et al.* 1976).

Generally, the use of binary data in environmental research can only be justified if it is difficult to obtain anything else, or if there is a declared lack of interest in the information which is lost by using binary data instead of, say, continuous data.

The attribute categories are not distinct, they depend to some extent on the sampling and coding procedure, and data in one form can be converted to another. In general, the conversion of continuous attributes (or discrete attributes that can be treated as continuous) to binary attributes is usually unsatisfactory. If a normally-distributed variate is divided into two sections along the mean, all objects on either side of the mean would have identical binary scores, however far from the mean. Rubin (1967) drew attention to a difficulty which arises when a continuous attribute is chopped into a set of intervals each of which is scored as a separate attribute-state. He used the example of age, which could be changed to a discrete attribute of 8 states thus:

(1)	0–9	(5)	40–49
(2)	10–19	(6)	50–59
(3)	20–29	(7)	60–69
(4)	30–39	(8)	over 69

The obvious difficulty when using this approach is that two persons aged 29 and 30 will be regarded as dissimilar, whereas two persons of 30 and 39 will be regarded as similar. He suggested that in the calculation of similarities, the problem could be overcome by having the user specify two different criteria: (a) an interval, expressed in the units of a variate, such that two objects which have a smaller difference on that variate will be considered to match (so that a one will be added to the number of matches when computing the similarity coefficient); (b) a second interval, expressed in the units of a variate such that two objects which have a larger difference on this variate will be considered not to match (and a zero will be added to the number of matches). For differences which lie between these two specified values he suggested using linear interpolation. This method has the advantage of avoiding sharp discontinuities in a similarity coefficient when changes in the data are slight, and the burden on the user to provide values for the two intervals is no greater than that of breaking a continuous variate into arbitrary intervals. A similar method could be used for ordered multistate data.

Also, in the calculation of similarity coefficients, there can be problems with mixed attributes. Rubin (1967) drew attention to difficulties which occur when different discrete attributes have different numbers of states. If we give equal weight to a binary attribute and to an attribute with four states, then, on average, the binary attribute will contribute matches more often to a similarity coefficient than will the four-state attribute. Furthermore, if we fragment the states of an attribute sufficiently, a match between two objects will become more and more rare, and the attribute will become useless in the analysis. Weighting attributes on the number of states is not the solution, since then an occasional match might completely dominate a similarity coefficient. The problem really lies in the original choice of states for each attribute. If there is too much variation in the number of states from attribute to attribute, then the many-state attributes will not play as important a part in a classification procedure as will the few-state attributes.

On the other hand, if one chooses states so that most objects assume only one or two of the states chosen, then one has thrown away information which could have increased our knowledge of the structure of the data set. Rubin suggested that the ideal solution might be to have an equal number of states for each attribute, and approximately equal frequencies for each state. This problem is not peculiar to the calculation of similarity coefficients. Once an attribute has been fragmented into too many states, or lumped into too few, information has been lost and is irrecoverable, no matter what the type of analysis. The particular problem for classification is that a randomly-chosen binary attribute may be more important in determining structure than several many-state attributes, even if the latter exhibit a high degree of structure when considered by themselves.

3.2 Selection of objects and attributes

The selection of both objects and attributes needs careful thought. A first problem is to decide whether to analyse a sample or the universe of objects. The answer will depend upon the nature of the problem, and usually the universe of objects is so large that it is convenient to handle only a sample. If a sample is taken, it may or may not be random. Random sampling maximizes the probability that the sample will reflect the general properties of the universe of objects, and minimizes bias. The theory and methods of sampling are too extensive to be dealt with in detail here. Two standard works are Yates (1949) and Cochran (1963), and there is a chapter on sampling in Webster (1977). Repetition can be important in exploratory work. If further sampling from the population is not possible, it can be useful to split the sample into several subsamples. This procedure can yield more information than does a single analysis, particularly with reference to the stability of the results, especially if the number of objects is large.

In environmental studies, there may or may not be *a priori* grounds for selecting attributes, and some attributes may be selected intuitively. There is likely to be some limitation on the types of attribute which can be used, due to practical difficulties in their measurement. The attributes chosen should be adequate to cover the domain of interest, and should be as widely representative as possible of variation in that domain. They may be chosen because of their theoretical importance, their relevance to some hypothesis, or their relevance to a specific problem. It can be useful to have 'marker' attributes (and objects) to cross-reference with other studies. Practical considerations are likely to limit the number of attributes used. Apart from the fact that a large number of attributes will result in more computer time (and storage space) being required in the analysis of the data, mathematical and statistical properties of the data matrix must be considered. The larger the number of attributes, the more likely it will be that many of the attributes will be highly correlated, and will thus contribute no useful information. In analysing the eigenstructure of a matrix, the larger the matrix, the less stable are the eigenvalues. If there are too many binary attributes, values of similarity coefficients will depend on accidental matches or 'noise'.

Although it is important not to include too many attributes, the researcher will be concerned that he/she may use too few, and thus omit an important attribute. An empirical approach to assist the user in deciding whether or not to add additional attributes would be to perform an analysis on the original set of attributes, add more, and repeat the analysis. If the results are similar, the classification is stable and the additional attributes are not required. A better approach would be to make a preliminary study of the proposed attributes by principal component or factor analysis. In environmental studies, observed attributes are often manifestations of a smaller number of factors of which the observer may be unaware. The attributes which the investigator thinks are important may be informative, but they may not be the ones which are correlated with real structure (e.g. see Muir 1962). Principal component or factor analysis can assist in the finding of important attributes in the underlying factors.

However the attributes are chosen, it is necessary to recognize that certain types of relationship among the attributes can cause problems in subsequent analysis. For example, secondary attributes which are derived from other attributes by arithmetical operations should normally be avoided. Such attributes may, in factor analysis (in the general sense), produce factors that are functions of the arithmetical operations on the data and not of the data themselves (Rummell 1970). Correlated attributes are another possible source of problems (Jardine and Sibson 1971; Sneath and Sokal 1973; Williams 1976). For example, Pythagorean distance calculated using correlated attributes without correction for the amount of correlation, will be over- or under-estimated by an amount which depends on the sign and magnitude of the correlation coefficient (Blackith and Reyment 1971 p 36). A group of correlated attributes can be replaced by one of its members, or by a new attribute constructed from all of them. An example of the latter is in principal component analysis, which can be used to orthogonalize the data before distances are calculated. A group of highly-correlated attributes may well represent some single underlying phenomenon, and if all the attributes are included, this amounts to a weighting of that phenomenon. The purpose of factor analysis methods (see Chapter 4) is to search for such underlying phenomena.

Certain statistical properties of the data can cause problems in subsequent analysis. If there are extreme frequencies, for example if most of the objects score zero on an attribute and only a few objects (say less than 10%) have non-zero scores, then relationships between this and other attributes hinge on only a few non-zero values. Restriction of range is a problem which can arise when a product–moment correlation matrix is factored. If an attribute is measured so that more than one object can take the same value, the range of the product–moment correlation coefficient with other attributes may be restricted to less than $-1 \leqslant r \leqslant 1$. The restriction will depend on the distributions of the attributes being correlated and the number of identical values in each direction. If one of the distributions is skewed in a direction opposite to the other, the positive range of the correlation coefficient will be restricted, if both distributions are skewed in the same direction, the negative range will be reduced (Carroll 1961).

Other criteria being equal, the researcher may wish to select those attributes that are most continuous or have the larger number of scale values, for example, preferring interval-scaled attributes to those measured on only four values, attributes with non-tied rankings over those with tied rankings, attributes measured on three values over binary attributes. If the attributes are similarly measured, it may be preferable to select those attributes with the greatest frequencies for the central scale values. For attributes with only three or four values, or for binary attributes, the researcher may wish to select those with the most rectangular distributions (Rummell 1970). If selection of attributes is not possible, some form of distributional transformation of the data may be necessary (section 3.5).

Marriott (1974) discussed the problem of binary attributes in cluster analysis. He noted that the selection of attributes, important in any multivariate procedure, is paramount in the case of binary attributes. If there is any inherent structure in the data, it should reveal itself in the dependencies between the variates. As a corollary, any variates that are non-independent for reasons not connected with an underlying grouping should be excluded from the analysis. For binary attributes, the problem of deciding if there is more than one group, and if so how many and how they should be divided, is not easy. The justification for multimodality as a criterion is less clear in the case of binary variates than in that of continuous variates. In general, multimodality depends in a complicated way on the probabilities associated with each dichotomy in a 2^p contingency table for p binary attributes.

3.3 Weighting of the attributes

The question of whether, or how, to weight data is an important problem in taxonomy, and specialists in different groups of organisms will have their own ideas about the importance of different attributes. A similar problem occurs in environmental research. One solution to this problem is to state the basis of weighting, so that the reader may judge the merits of the case (Clifford and Stephenson 1975). Sneath and Sokal (1973) considered that equal weighting is desirable; it can be defended on several independent grounds, and is probably the only practical solution.

Jardine and Sibson (1971) concluded that certain kinds of weighting which taxonomists use intuitively are, in fact, incorporated in the calculation of K-dissimilarity, while certain of the other kinds of weighting and correlation which taxonomists have discussed were shown to be relevant to the selection of attributes, rather than to the calculation of dissimilarity and analysis of dissimilarity coefficients once attributes have been selected.

In the raw data, it is likely that several attributes might reflect some single underlying feature, and the use of such attributes is an implicit weighting of that feature. One way of avoiding this, for example in cluster analysis and regression, is to use component values after principal component analysis, provided that the data are suitable for this method.

3.4 Possible structure in the data

As an important part of pattern analysis is the search for discontinuities in the data, it is important to think about what types of pattern may be present in the data.

The first possibility is that there is only one group and all the objects belong to it (e.g. Figure 3.1). We may or may not know the nature of the distribution, but we cannot assume that it is normal. The data may contain more than one group. If there are real discontinuities, the groups will be separate and distinct. This situation presents no problems, the problems occur if the tails of the frequency distributions overlap (e.g. Figure 3.2). Examples with a greater degree of overlap can be visualized, with the centres of the distributions moving closer together. In the extreme, this leads to a complex distribution which may appear to be unimodal (Figure 3.3).

This type of pattern is visualized by plant ecologists. Sorensen (1948) suggested that the various types of vegetation are often so insensibly merged as to form a sliding scale, but that in a limited area under investigation it can be considered to be homogeneous with as much approximation to that mathematical concept as can be found in nature. Whittaker (1970, 1972) developed the continuum theory and showed that one might expect each species to find a different niche on an environmental gradient (e.g. Figure 3.3). If the distribution patterns of species are completely continuous, it

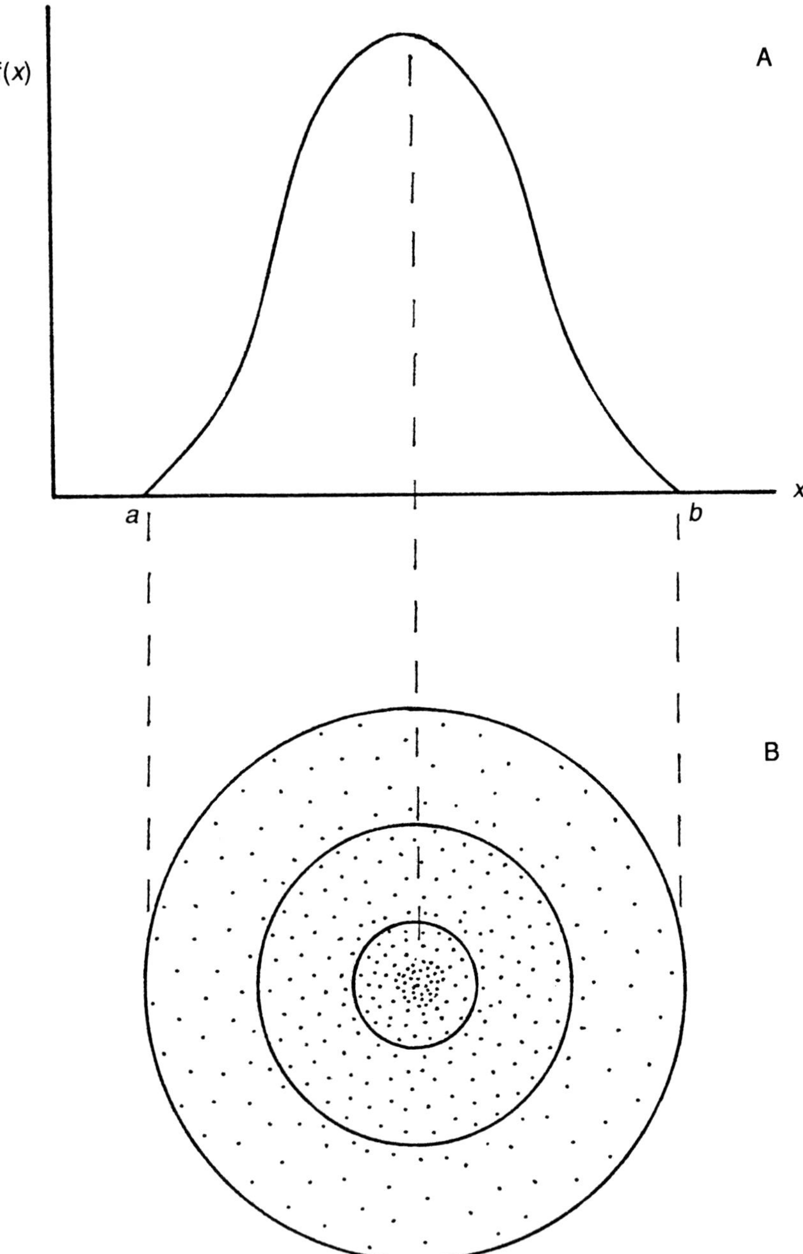

Figure 3.1 There is only one group, and all the objects belong to it. A is the graph of the continuous frequency (probability density) function $f(x)$ of the data. This function is zero outside some finite interval (a,b). For convenience, it is shown as univariate and symmetrical. B represents a slice through a bivariate version of A, at right angles to the paper, to show the density gradient

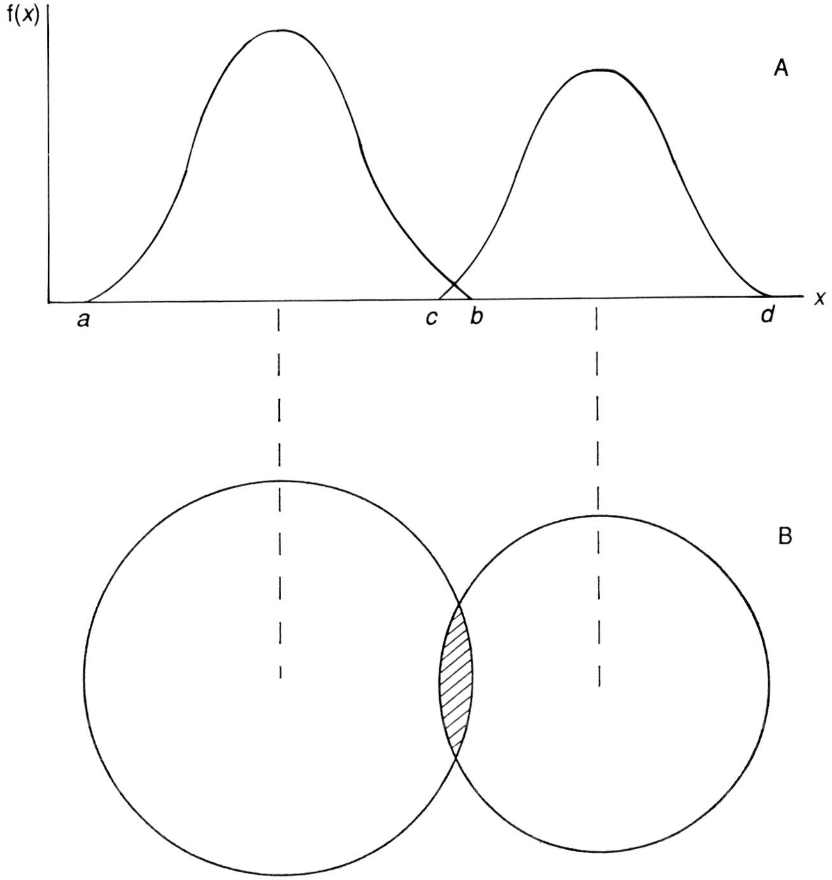

Figure 3.2 A bimodal sample with the tails of the frequency (probability density) distributions slightly overlapping

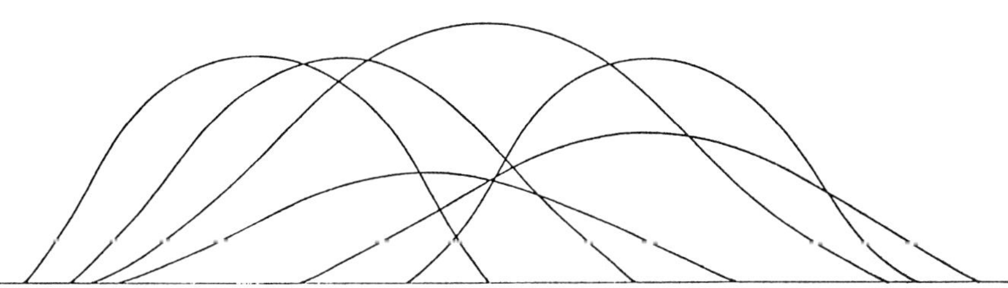

Figure 3.3 Complex overlapping distributions as visualized by some plant ecologists

becomes impossible to delineate communities or associations, and most difficult for the human brain to comprehend the data in totality. Where gradients are involved, ordination may be the best way of handling the data, although that, too, has its problems. The concept of such a unimodal continuum is not unique to ecology, examples can be found in other sciences (Clifford and Stephenson 1975). The possibility that such a continuum may be divided into groups on the basis of so-called homogeneous areas led to the development of a variety of methods for this purpose, usually based on some sort of variance constraint. These methods, and the problems associated with them, will be discussed later.

3.5 Scaling, standardization and transformation of the data

The raw data may not be uniform. In ecological studies, for example, this may be because some species are more abundant than others, or because attributes measured on different scales differ in both range and variability. The importance of such differences in the analysis of the data needs thinking about, particularly in relation to the objectives of the investigator. In one context, it may be useful to exclude rare species (cf. Barkham 1968). For some purposes, it may be judged that attributes should contribute equally regardless of their variation. On the other hand, the investigator may consider the variation itself to be an important feature to be retained.

We can recognize three ways in which the data may be modified: (1) Scaling; (2) Standardization; (3) Transformation.

Scaling

The simplest form of scaling is *translation*, which is the addition of a constant to, or its subtraction from, each value of an attribute. Geometrically, this procedure moves the origin of the measurement scale. Translation does not affect distance measures, but angular measures of resemblance are changed if the angle is measured at the origin. If a constant is added to each value of an attribute, the new mean is equal to the original mean plus the constant. The variance and standard deviation are unchanged.

Expansion is the multiplication, or division, of each value of an attribute by a constant. It corresponds to a change in units of measurement. In general, expansion changes both distance measures and angular measures of similarity, depending on the relative changes of the scales of different attributes. If all the original values of an attribute are multiplied by a constant, the new mean is equal to the original mean multiplied by the constant. The new variance is equal to the original variance multiplied by the constant squared, and the new standard deviation is equal to the original standard deviation multiplied by the constant.

Standardization

Standardization means that the value of each attribute for each object is expressed as a deviation from the mean of that attribute and divided by the standard deviation. This has the effect of reducing all attributes to unit standard deviation, and of reducing the magnitude of each attribute. Other methods are ranging (Sneath and Sokal 1973 p153) and rankits (Sokal and Rohlf 1969). In ranging, the smallest value for an attribute is subtracted from each value and the result is divided by the range. Thus, the smallest attribute score is set at zero and the largest at 1. If the attributes are normally distributed, the ranged values will be approximately a monotonic function of the standard deviations. A rankit is the average of the kth largest deviate ($k = 1$ to N) in a sample of N observations drawn at random from a normal distribution with a mean of zero and a variance of unity. This procedure has an effect not given by standardization or ranging, namely conversion of the frequency distribution of attribute scores so that they are approximately normally distributed.

There appears to have been no employment so far in numerical taxonomy of standardizations that equalize the variability while leaving gross size unchanged (Sneath and Sokal 1973). Standardization of attributes makes them dimensionless, and so renders them additive. It also reduces the range, so that a single attribute, or a small number of attributes, does not automatically dominate the analysis. The choice of the method of standardization is effectively a form of weighting (Williams 1971). In some numerical analyses, it may be difficult to decide whether or not to standardize an attribute. The information content (in the information theory sense) of normally-distributed continuous attributes is proportional to their variance. If the variances are made equal, then each attribute contributes equal information. Clearly, the effect of this needs careful thought, particularly with regard to contributions to the variances from sources such as random error (Sneath and Sokal 1973). In numerical taxonomy, standardization is rarely employed with presence–absence data. Its effect is to give greater weight to mismatches in attributes which are rare or very common. In a comparison of methods for the analysis of vegetation data, Williams *et al.* (1966) found that standardization of the data before the calculation of squared Euclidean distance was, if anything, disadvantageous as judged by other numerical and ecological criteria.

Transformation

The term 'transformation' is used of methods which seek to change the shape of the frequency distribution of the data, usually in the hope of obtaining an approximately normal distribution. In univariate statistics, for example, transformations may be used to satisfy the theoretical requirements of the analysis of variance. Classical multivariate theory has been based largely on the multivariate normal distribution, and although in some multivariate methods multivariate normality is not essential, except for the sampling theory (e.g. in canonical correlation), not much is known about the robustness of the methods and the effects of large departures from normality. Some methods are sensitive to non-normality, for example cluster methods based on the assumption of a mixture of multivariate normal distributions.

Various types of transformation have been used (see e.g. Rummell 1970; Sneath and Sokal 1973; Clifford and Stephenson 1975). Southwood (1966) gave a detailed discussion of transformations which may be useful in ecology, and which are of general use. Units of measurement are generally arbitrary. For example, extractable ions are conveniently expressed as milliequivalents per 100 grams, but this is only a convention and there is no *a priori* reason for this particular unit. Particle size may be given as the diameter in millimetres, but this is also arbitrary, and a transformed scale may be no less valid and may prove more useful. A commonly-used transformation is to take the logarithm (usually to base 10) of the value. In order to be logarithmically transformed, data should have a natural origin. It may therefore be necessary to add a constant to each value first, the constant being chosen to make the smallest value of an attribute 0.5 or 1. The logarithmic transformation constricts the intervals of the data as the values increase in size, consequently it is useful with distributions which are skewed to the right. The right tail is drawn in and the values to the left of the distribution are moved away from the mean, tending to normalize the distribution.

Occasionally, the logarithmic transformation is too powerful, it may pull the right tail in too far and create outliers on the left side of the distribution. In such a case, a weaker left skew transformation such as the square root or reciprocal may be useful. For the square root transformation, the values must all be non-negative, or must be made to be. For the reciprocal transformation, the values must be all positive as in the logarithmic transformation. The above transformations may also reduce heteroscedasticity, the logarithmic transformations being particularly useful for this purpose.

Where the attributes all have the same units, e.g. lengths, the effect of logarithmically transforming all of them is to give the same variance to measurements with the same proportional variability. At the same time, it is not a method of standardization, as attributes that are relatively more variable will have greater variances when transformed. The logarithmic transformation is particularly suitable for counts which range from low to very high values (Marriott 1974).

If the data are skewed to the left, the appropriate normalizing transformation is likely to be one of the powers of X greater than unity, the strength of the transformation increases with increasing value of the power of X used. However, the larger the power used, the greater the loss of information from the original data. In these transformations the data must be non-negative, as power transformations require the data to have a natural origin (McNeil 1977). Other transformations which can be useful in normalizing data which are highly skewed to the left are $\log [X/(1 - X)]$ for $0 < X < 1$, and $0.5 \log [(1 + X)/(1 - X)]$ for $0 \leqslant X < 1$, the latter being more powerful (Rummell 1970). The former is an example of a folded transformation, and is particularly useful with proportions or data that can be converted to proportions (see McNeil 1977). Another strong transformation is arcsin X for $0 \leqslant X \leqslant 1$. Arcsin may be used for proportions whenever binomial distributions may be assumed (Sneath and Sokal 1973). Smartt *et al.* (1976) examined the effect of the arcsin transformation on vegetation data.

If the distribution is J-shaped (or reverse J-shaped), no simple, non-grouping, transformation will normalize the distribution. However, transformations may still be applied to eliminate the outliers, or reduce their size.

In principal component analysis and factor analysis, normal-standardizing (or ranking as a first approximation) is recommended for variates with substantially-skewed distributions. In normal-standardizing multistate data, a unit normal distribution is first divided into segments with areas proportional to the frequencies of the original states. The mean method of normalizing puts each z at the mean of the corresponding segment, where z is the standard-score distance from the mean of the whole distribution to the mean of the segment. The z values are then used to calculate new attribute scores which are used to compute the correlation matrix. The procedure is described in detail in Cureton and d'Agostino (1983 p127). Normal-standardizing scores reduces artificial non-linear regressions due to differences in direction of skew. However, if the new scores are not linearly related there is nothing more that can be done.

Andrews *et al.* (1971) discussed the problem of transformations in which the transformed variates are functions of the original variates collectively rather than separately, and suggested some techniques which might be useful.

If the data are species presence–absence, some species will be more abundant, others less so. Use of such data without some form of transformation can lead to abundance being over-emphasized, the results of the analysis being dominated by a few very large values. Data of this type can have important effects on distance measures, which are discussed in Chapter 11. In particular, if Pythagorean distance is used, the squaring of differences greatly emphasizes the difference between rare and abundant species. The same holds for clustering methods using the sum of squares criterion. The Bray–Curtis measure is also sensitive to large values and its use emphasizes dominance. Whether this is important or not depends on the objectives of the user, but it is a factor which must be borne in mind. If there are occasional very large values, it may be better to reduce the importance of abundance, e.g. by a logarithmic transformation. Alternatively, one could use a measure, such as the Canberra measure, which is insensitive to occasional large values. This could be used with or without a mild transformation such as the square root. On the other hand, if the data have a high diversity, with several species more or less equally abundant, it may be better to concentrate on the abundant species at the expense of the rarer ones (Clifford and Stephenson 1975).

Clifford and Stephenson (1975) stated that it remains uncertain whether the transformation required to produce normality of data is also the transformation which will produce optimal ecological 'sense', and that optimal ecological classificatory 'sense' is generally obtained by using a weaker transformation than that required to transform data to normality. In any multivariate analysis, careful thought needs to be given to the nature of the data and how this relates to the methods to be used. Despite any constraints used, attributes will not, in general, contribute equally to the final classification. An attribute which varies little over the population has little or no discriminating power.

It should be remembered that transformations can easily alter the information content of the data, even to the extent that it changes the outcome of the analysis (Orloci 1978 p9).

3.6   Attributes of constant sum

The commonest example of attributes of constant sum (called ipsative measures) is the case of three attributes, commonly found in geology and pedology. An example is where soil particle size must belong to either sand, silt or clay. Expressed as a percentage, these must, by definition, add up to 100% for a given soil sample. Data of this type may be plotted on triangular co-ordinate paper, which projects the points onto a plane. The properties of such data were discussed in detail by Koch and Link (1971). It is clear that only two of the three attributes need to be known, because the third is 100% minus the sum of the other two. With p such attributes, $p - 1$ attributes at most can be independent (see section 2.1).

Such data are said to form a closed array, and they raise the problem of whether the covariances contained in the array could have been generated by the closure of an open array consisting of uncorrelated attributes. This problem was discussed in detail by Chayes and Kruskal (1966), who showed that the expected values of the induced correlations among attributes in a closed array can be calculated from the observed covariances. If the observed correlations are different from these expected values, then the conclusion is that relationships exist among the attributes in the closed array that are not due merely to correlations induced by the closure of the array.

Section 2

Numerical methods

CHAPTER 4

INTRODUCTION TO ORDINATION METHODS

Initially, our data matrix $\mathbf{X}$ ($N \times p$) defines the positions of the objects in a p-dimensional attribute space. Or it may be that we begin not with a data matrix but with some sort of dissimilarity matrix. Ordination procedures aim to arrange the objects along one or a few new dimensions so as to preserve as much as possible of the original information, i.e. to preserve the relationships between the objects as accurately as possible. The reduction in dimensionality makes the data easier to handle mathematically: (1) it makes graphical representation easier; (2) it removes difficulties which might arise from attributes which are linearly related, or nearly so; (3) the axes resulting from the reduction in dimensionality may lend themselves to reification, i.e. the interpretation of the mathematics in terms of the original problem, and may give a useful insight into the structure of the data. If there are 'natural' groups, i.e. groups which are separated by discontinuities in multidimensional space or by regions of the space containing sparse points which represent the tails of overlapping distributions and/or 'noise' data, this should be apparent in the results of the ordination. If there are no such groups, ordination may still throw some light on the relationships between the objects, and it is particularly useful if the objects are distributed along gradients. Special problems occur with the particular types of distribution of plant species along environmental gradients. Ordination may also show if a clustering method has been used for data to which it is not suited.

In most ordination procedures there is no need for *a priori* information about structure in the data. However, in canonical variate analysis *a priori* information about groups of objects is required.

In this section, we shall examine a selection of ordination procedures, including the more important and widely-used ones. It is convenient to begin with 'factor analysis' procedures, first the R-mode (principal component analysis and 'true' factor analysis), then the Q-mode (factor analysis, using Imbrie's method, principal co-ordinate analysis, and correspondence analysis), and then to mention canonical correlation and canonical variate analysis. Imbrie's method is not widely used, but it is included partly for completeness and partly because it may prove to have wider applications.

R-mode methods: principal component analysis and factor analysis

It is useful at this stage, before going on to discuss the methods in detail, to outline the important differences between principal component analysis and factor analysis. These methods have in common the fact that they aim to find a small number of hypothetical attributes (components or factors) which contain the essential information expressed in a larger number of observed attributes. Hence, the dimensionality of the data is reduced by utilizing interdependencies. Factor analysis was originally developed for the analysis of the scores obtained by individuals on batteries of psychological tests, and it still has many applications in that field, although its use has spread to other fields, notably geography (e.g. see Mather 1976) and geology (e.g. see Jöreskog *et al.* 1976). Originally, the term 'factor analysis' included principal component analysis, but there are important differences which make it more convenient to distinguish between them. The methods are based on mathematical models which present simplified, but not exact, representations of the data based on certain assumptions. The differences between the techniques are largely connected with differences in these assumptions.

Cattell (1965) pointed out that in principal component analysis, except for the measurement errors, any attribute would correlate perfectly with itself. With this model, the number of components will, in general, equal the number of attributes, although some may make a negligible contribution and can be ignored. With the factor analysis model, on the other hand, there will usually be a limited number of common factors. A growing plant could well prove to be influenced by most things in the universe, but it would be absurd to expect all of those influences to be substantially represented in any small set of attributes. Consequently, the correlation of any attribute with itself due to the common factors represented among those attributes is likely to be decidedly short of unity. Part of its variance must be set aside to be accounted for by common factors which will only appear when an extremely large number of attributes has been used to form a correlation matrix.

Thus, in the principal component analysis model, the assumption is that a small set of attributes sampled at a particular time accounts for all the variance of each attribute in common factors peculiar to that restricted set of attributes, while in the factor analysis model some variance is left open to later resolution. Cattell (1965) considered it 'wildly unlikely' that a small sample of attributes will actually represent the real common influences required to account for all the variance, and that the practice of putting ones into the diagonal of a correlation matrix 'perpetuates a hoax', for it actually 'drags in' all the specific factor and error variance to inflate 'specious, incorrect, common factors'. He considered that this practice has included in the common factor space much variance about which we know nothing more than is known about specific factors. This quickly becomes evident in practice when we try to match factors from one piece of research with those from another. If one is testing a hypothesis that a factor in one set of observations is the same as a factor in another set, and the observations have some, but not all, attributes in common, the common factors of one set of observations cannot possibly be made accurately to match those in a second set, even theoretically, because they derive from a different set of specific attribute factors in each set. In the factor analysis model, the specific factor variance is set aside as variance needing further investigation.

Mathematically, we can look at the method in two ways. Firstly, we have a model for the data matrix or 'fixed case', in which we are interested only in a given set of data. Secondly, we have a model for the variances and covariances, the 'random case', in which the data are regarded as a random sample from some population. In this case, we wish to make inferences about the population from an analysis of the sample. This approach involves questions such as the nature of the multivariate distribution of the variates and also sampling theory.

The model for the data matrix (fixed case) is

$$\mathbf{X} = \mathbf{FH'} + \mathbf{E} \qquad (4.1)$$

where $\mathbf{X}$ ($N \times p$) is the data matrix, $\mathbf{F}$ ($N \times q$) is a matrix of factor scores, $\mathbf{H}$ ($p \times q$) is a matrix of factor 'loadings' (weightings), and $\mathbf{E}$ ($N \times p$) is a matrix of residuals or error terms which are assumed to be uncorrelated with the factors (Jöreskog *et al.* 1976; Mather 1976). The number of factors q is almost always less than the number of attributes p. For any object in the data matrix (the vectors are row vectors):

$$\mathbf{x'} = \mathbf{f'H'} + \mathbf{e'} \qquad (4.2)$$

The product $\mathbf{f'H'}$ is the best linear estimate of $\mathbf{x'}$. For any object and any attribute, this estimate is called the 'common part', because it has something in common with the other attributes via the factors. The residual term is taken to be the sum of two uncorrelated parts, the 'specific part' and the measurement error.

In the model for the variances and covariances, the total variance of a variate is assumed to be made up of two parts, the 'common variance' or 'communality', and the residual or 'unique' variance. The unique variance itself can be divided into the 'specific' variance and the error variance. The communality represents the portion of the variance of a variate which is in common with the other variates, and is involved in the covariances. Thus, the communalities are large for variates showing large intercorrelations. The specific variance denotes the variances specific to a variate, and the error variance is due solely to measurement error. If Σ ($p \times p$) is the population covariance matrix, then

$$\Sigma = \mathbf{H}\mathbf{\Phi}\mathbf{H}' + \mathbf{\Psi} \tag{4.3}$$

where $\mathbf{\Phi}$ ($q \times q$) is the covariance matrix for the factors and $\mathbf{\Psi}$ ($p \times p$) is the residual covariance matrix.

The fundamental differences between principal component analysis and factor analysis lie in the ways in which the factors are defined, and in the assumptions about the nature of the residuals. In principal component analysis, the factors (components) are determined with maximum variance constraint. In factor analysis, the factors are defined to account maximally for the intercorrelations of the attributes. In principal component analysis, the residual terms are assumed to be small, so that

$$\mathbf{X} \simeq \mathbf{F}\mathbf{H}' \tag{4.4}$$

and a large part of the total variance of an attribute is assumed to be important. In factor analysis, there is no such assumption, and only that part of an attribute is used that participates in correlations with other attributes. In both methods, the residuals are assumed to be uncorrelated with the factors. In principal component analysis, there is no assumption about correlations among the residuals, whereas in factor analysis the residuals are assumed to be uncorrelated among themselves.

With regard to the covariance matrix, in principal component analysis the residual covariance matrix $\mathbf{\Psi}$ is assumed to be small, so that

$$\Sigma \simeq \mathbf{H}\mathbf{\Phi}\mathbf{H}' \tag{4.5}$$

In factor analysis the residual variances are assumed to be uncorrelated among themselves, so that $\mathbf{\Psi}$ is a diagonal matrix of unique variances. Hence, the off-diagonal elements of Σ are the same as those of $\mathbf{H}\mathbf{\Phi}\mathbf{H}'$, and the diagonal elements of the latter are the communalities of the attributes. Thus, in principal component analysis the aim is to reproduce all the elements of Σ, in factor analysis the concern is with its off-diagonal elements, i.e. the covariances (Jöreskog *et al.* 1976). Since factor analysis is concerned with extracting common information on covariation, the method will fail if adequate common information does not exist.

Principal component analysis is probably the best-known and most widely used ordination technique. As with other ordination procedures, the positions of the objects can be plotted on pairs of rectangular Cartesian component axes. Such plots will show discontinuities if they exist in the data (e.g. see Blakith and Reyment 1971), but it must be remembered that any such two-dimensional representation is distorted in that other dimensions are not included. Gower and Ross (1969) showed how such distortions can be illustrated by superimposing the minimum spanning tree (see Chapter 15) of the points in the total dimensionality on to their representation in the reduced space. There has been much discussion about the use of principal component analysis in plant ecology, because of problems caused by the nature of plant species distributions along environmental gradients. Too many papers have been written for detailed discussion here. See, for example, Bray and Curtis (1957), Austin and Orloci (1966), Beals (1973), Noy-Meir (1973), Whittaker (1973), Dale (1975), Austin (1976), Orloci (1978).

Factor axes (and component axes) may be rotated to determinable positions in which they are not necessarily, or even generally, orthogonal. Sneath and Sokal (1973) considered that this makes scientific sense in that the factors underlying the covariation pattern of the attributes in nature are themselves undoubtedly correlated, but they pointed out that there are problems.

Blackith and Reyment (1971) stated that it is very hard to discuss factor analysis without generating more heat than light; it is the most controversial of the multivariate methods. Factor analysis was proposed originally as a model for a well-defined problem in education psychology, but it acquired a bad reputation among mathematicians and was largely ignored outside the field of psychology, where it still finds most of its applications. The method and criticisms were discussed by Cattell (1965), Blackith and Reyment (1971), Lawley and Maxwell (1971), and Marriott (1974). Clifford and Stephenson (1975) went so far as to state 'It is likely that in the future factor analysis will play an increasingly important role in ecological studies'. On the other hand, Gower (1967b) considered it doubtful if factor analysis really is a helpful way of viewing biological data, and Blackith and Reyment (1971) asked: 'Could it not be that factor analysis has persisted precisely because, to a considerable extent, it allows the experimenter to impose his preconceived ideas on the raw data?'

While it is true that the statistical and mathematical foundations of factor analysis were poorly expressed and its aims were not widely understood, it should also be noted that over the past few years many theoretical and computational developments have taken place which put the methods on a sounder footing. The approach now deserves some serious consideration, although it should be approached with caution and with a clear understanding of what is involved. A very readable account of the aims and methods is given by Mather (1976). Jöreskog (1977) gave a rather mathematical account of least-squares and maximum-likelihood methods, and Harman (1977) gave an account of the Minres method.

The structure of correlation matrices

Correlation matrices are important in principal component and R-mode factor analysis. The most widely-used measure of correlation between two attributes is Pearson's product-moment correlation coefficient. In calculating this coefficient for two attributes, the underlying distribution is assumed to be bivariate normal. Although there is some evidence which suggests that the product-moment correlation coefficient can be used if the parent population is moderately skewed, if the distribution is J- or U-shaped it may be better to use a non-parametric measure of correlation, such as Spearman's rank correlation coefficient, otherwise the use of the product-moment coefficient could well mask, rather than reveal, relationships (Mather 1976 p242).

Carroll (1961) showed that if an attribute is measured in such a way that more than one object can take the same value, the usual range of the product-moment correlation coefficient $(-1 \leqslant r \leqslant 1)$ may be restricted. The restriction will depend on the distributions of the two attributes being correlated and on the number of identical values in each distribution. If the distributions are skewed in opposite directions, the positive range of the correlation coefficient will be restricted; if both distributions are skewed in the same direction, the negative range will be reduced. If it is impossible to transform the attributes to approximate normality, a non-parametric coefficient may be preferable.

In principal component analysis and factor analysis, normal-standardizing (or ranking as a first approximation) is recommended for variates with substantially-skewed distributions (see section 3.5).

The product-moment correlation coefficient also assumes that a zero correlation measures statistical independence of the attributes (Kendall and Stuart 1973 p300). If the distributions of two attributes are heteroscedastic, the correlation coefficient is not a measure of statistical independence. Homoscedastic distributions can only be guaranteed if the marginal distributions of the attributes are normal. Another problem which may arise is the presence in the data matrix of an attribute which exerts an influence over many of the other attributes. The correlations then reflect the influence of this attribute rather than simple bivariate relationships. Mather (1976 p243) suggested that in such circumstances the use of partial correlation coefficients may be justified. Their use is equivalent to treating the attribute to be partialled out as having the status of a factor.

If it is considered to be essential to compute correlations between binary attributes, there are usually two possibilities. If the underlying distribution is normal and data which are really continuous have been scored as binary (1,0) for practical convenience, the tetrachoric correlation coefficient should be used (Kendall and Stuart 1973 p316). On the other hand, if a basic dichotomy in the attributes is believed to exist, then the fourfold point coefficient is more likely to be appropriate (Sokal and Sneath 1963, p155). However, a correlation matrix produced in this way may be non-Gramian, its rank may be increased, and in factor analysis this is likely to lead to over-factoring and the possibility that some of the communalities may be inflated. Binary data lead naturally to the construction of contingency tables and similarity coefficients, and therefore to ordination by methods such as principal co-ordinate analysis and correspondence analysis or to clustering methods working directly with the similarity coefficients.

Procedures for dealing with missing values were discussed by Cureton and d'Agostino (1983 p135).

Q-mode methods

Factor or component scores from an R-mode analysis can be plotted on pairs of rectangular Cartesian axes, and these scores provide a means of describing inter-object relationships. However, covariance is not the only basis, and may not necessarily be the best, on which to examine such relationships. The basis of Q-mode methods is in a definition of inter-object similarity, or association, and the calculation of a similarity matrix of order $N \times N$. Given such a matrix, the steps in the procedures are analogous to those in R-mode analysis, but the interpretation of the results is different.

Q-mode factor analysis was first applied in geology by Imbrie, and we can use his method as an example of this technique. Here, similarities are defined with respect to properties of constituents, for example mineral species, which make up a rock or sediment. Gower's method of principal co-ordinate analysis is based on the use of a distance or dissimilarity matrix. The ability to ordinate a set of objects given only their dissimilarities can be useful in environmental studies, and there are some circumstances in which a particular dissimilarity measure might be preferred. For example, one might wish to emphasize dominance and thus use the Bray–Curtis measure, or, perhaps, be more concerned with relative properties and use the Canberra metric (Clifford and Stephenson 1975). Principal co-ordinate analysis is particularly useful when there are missing values or missing variates. In such a case, a correlation type of similarity measure is reasonably robust and reliable, whereas replacing the missing values by estimates or guesses is not very satisfactory (Marriott 1974).

Benzécri's method of correspondence analysis, based on the use of a contingency table for assessing similarities, is rather more complex than the previous methods because it aims to obtain, simultaneously, both Q-mode and R-mode factor loadings, and the similarity must be defined jointly and symmetrically. This method permits the simultaneous presentation of objects and attributes as points on the same pair of co-ordinate axes, and mutual dependencies can be interpreted.

Allied to the above methods are two other multivariate techniques which investigate relationships in multidimensional space, but which operate on data which are already grouped either on the basis of objects (canonical variate analysis also known as canonical analysis of discriminance) or variates (canonical correlation analysis). In canonical variate analysis, the relationships of the groups to each other in multidimensional space are investigated. As with the above procedures, the canonical variate space usually has a lower dimensionality than the original attribute-space. Strictly speaking, canonical variate analysis is not one of the general pattern analysis methods, and will not be dealt with in detail. It is mentioned briefly here because it can be useful if it is desired to examine the relationships among groups in an existing classification, or to test the allocation of new individuals to existing groups. It requires a non-singular within-groups covariance matrix which should be constant over all groups, and the data must be substantially multivariate normal. In canonical correlation analysis, the aim is to select pairs of maximally-correlated linear functions from the two batteries of variates. Again, this reduces the dimensionality. Gittins (1979) has given a comprehensive account of canonical correlation analysis and its ecological applications.

CHAPTER 5

PRINCIPAL COMPONENT ANALYSIS

The theory of principal component analysis is dealt with in some detail for two reasons. Firstly, the method is useful, and widely used, in a variety of studies, and some understanding of the theory is necessary in order to appreciate the usefulness, and limitations, of the method. Secondly, it provides a useful introduction to factor analysis type methods, and it is useful to be able to draw analogies with it in discussing other methods.

Theory: The eigenvalue decomposition of a covariance or correlation matrix

Suppose we have a sample of N independent observation vectors of p variates, giving the $N \times p$ data matrix $\mathbf{X}$. The variates may be correlated, and any interrelationships will be linear. It is preferable for the data matrix to be of full rank, and not to include variates which are linear compounds of the other variates, as such compounds can obscure whatever latent structure might be present. The population covariance matrix Σ is assumed to be positive definite, i.e. of full rank ($r = p$), and the eigenvalues are always real, distinct, and positive. In degenerate cases, some of the eigenvalues may be equal. If the $p \times p$ covariance matrix is positive semidefinite (singular) of rank r, it contains r positive eigenvalues and $p - r$ zero eigenvalues. As long as the population eigenvalues are distinct, the sample eigenvalues and eigenvectors are maximum likelihood estimators, and in most applications it can be assumed that the eigenvalues are different (Anderson 1958). Our estimate of Σ is the sample covariance matrix $\mathbf{S}$. Since we are interested only in variances and covariances, we can assume that the observations in the data matrix $\mathbf{X}$ are expressed as deviations from the sample means. The actual distribution of $\mathbf{X}$ is irrelevant, except for the structure of the covariance matrix, but interpretation of the components is easier if the data are multivariate normally distributed.

Components are normalized linear combinations of the variates, the first component having maximum variance and successive, uncorrelated (orthogonal) components having decreasing variances. In matrix notation,

43

the $N \times r$ component matrix $\mathbf{Y}$ is given by

$$\mathbf{Y} = \mathbf{XA} \qquad (5.1)$$

where $\mathbf{A}$ $(p \times r)$ is a matrix of weighting coefficients.

The covariance matrix of $\mathbf{Y}$, let it be $\mathbf{D}$, is given by

$$\mathbf{D} = \mathbf{A'SA} \qquad (5.2)$$

(Searle 1966; Morrison 1967; Seal 1968). For the transformation in (5.1) to be orthogonal, the off-diagonal elements of $\mathbf{D}$ must be zero. The statistical problem is to find a unique $\mathbf{A}$ such that the diagonal elements d_j are the maximized variances of the vectors $\mathbf{y}_j$, subject to the constraint $\mathbf{a}_j'\mathbf{a}_j = 1$. This constraint simply provides a scaling to ensure the uniqueness of the coefficients. It can be shown (Anderson 1958; Morrison 1967; Press 1972) that the vectors $\mathbf{a}_j$ must satisfy

$$(\mathbf{S} - d_j\mathbf{I})\mathbf{a}_j = 0 \qquad (5.3)$$

i.e.

$$\mathbf{Sa}_j = d_j\mathbf{a}_j$$

which represents a set of r homogeneous linear equations. If such a set of equations is to have a non-zero solution, the determinant of the coefficients must be zero, i.e.

$$|\mathbf{S} - d_j\mathbf{I}| = 0 \qquad (5.4)$$

which is the characteristic equation. The vectors $\mathbf{a}_j$ are the eigenvectors (latent or characteristic vectors) of matrix $\mathbf{S}$, and the scalars d_j are the corresponding eigenvalues (latent or characteristic roots). For maximum variance in the first component we use the largest eigenvalue. For the second component we need the maximum variance uncorrelated with the first, i.e. we need the orthogonality constraint $\mathbf{a}_i'\mathbf{a}_j = 0$ $(i \neq j)$. This variance is provided by the second eigenvalue, and the coefficients are the elements of the second eigenvector.

Matrix $\mathbf{A}$ is orthogonal (and the eigenvectors are orthonormal, see below), so that the transformation in equation (5.1) is distance- and angle-preserving (Anderson 1958 p338). The elements of $\mathbf{A}$ show the relative importance of each original variate in the new, derived component. Such an orthogonal transformation leaves invariant the generalized variance of $\mathbf{X}$, and the sum of the variances of the components is equal to the sum of the variances of the original variates (the sum of the eigenvalues of a matrix is equal to the trace of the matrix). It follows that the percentage of the total variance accounted for by each component is calculated easily.

The above theory has been developed for the covariance matrix, and this application is most suitable if all the original variates have been measured in the same units. Even if the variates are all on comparable scales, it is necessary to be careful in working with unstandardized data. For example, if the range of values obtained for one variate is greatly different from that of another variate, it clearly makes a good deal of difference whether we standardize the data or not. On the other hand, if the original variates are not in the same units, their linear compounds would have little meaning, and the rationale of maximizing $\mathbf{a'Sa}$ relative to $\mathbf{a'a}$ is questionable; in fact, the analysis will depend on the different units of measurement (Anderson 1958).

If the variates are not all expressed in the same units, we could standardize the data so that

$$z_{ij} = x_{ij}/s_j \tag{5.5}$$

where x_{ij} is an element of matrix $\mathbf{X}$ and is a deviation from the mean, and s_j is the standard deviation of the jth variate. In this case, the transformation is

$$\mathbf{U = ZB} \tag{5.6}$$

and the covariance matrix of $\mathbf{U}$, let it be $\mathbf{L}$, is

$$\mathbf{L = B'RB} \tag{5.7}$$

where $\mathbf{R}$ is the covariance matrix of $\mathbf{Z}$, and is the correlation matrix of $\mathbf{X}$.

The theory for principal component analysis based on the correlation matrix is essentially the same as that for the covariance matrix (see Cooley and Lohnes 1971). We compute the eigenvalues l_j and eigenvectors $\mathbf{b}_j$ of the correlation matrix. The sum of the eigenvalues will be the trace of $\mathbf{R}$ and will be equal to p. In analysing the correlation matrix, we are regarding the data as a battery of correlated variates, and the aim of the linear transformation in equation (5.1) is to obtain a battery of uncorrelated components with maximized variances, but in this case we maximize $\mathbf{b}_j'\mathbf{Rb}_j$ relative to $\mathbf{b}_j'\mathbf{b}_j$. Hence, the pattern of the eigenvector elements depends upon the correlation structure of the observations. In an analysis of the covariance matrix, the pattern of the eigenvector elements will depend upon the variance–covariance structure.

The use of the correlation matrix does present some theoretical problems. Press (1972) pointed out that the new variates are not really standardized relative to the population, and they do not have unit variance except asymptotically. In small or moderate-sized samples, the objective of finding linear functions of the original variates, or even their versions standardized relative to the population, with maximum variance, is not achieved. He recommended using the correlation matrix only for very large samples. If the elements of $\mathbf{X}$ are standardized relative to the sample before the components are extracted, the distribution theory associated with the sample principal components and their variances is mostly intractable.

Anderson (1963) showed that the sampling theory of components extracted from correlation matrices is much more complex than that of covariance components. He also commented that it seems a little incongruous to standardize the variates to unit variances and then to return to maximizing variances of linear combinations. Nevertheless, analysis of the correlation matrix has proved useful in practice, and Marriott (1975) stated that the theory, when applied to actual data, is at best approximate, and the error introduced by standardization is very unlikely to be important. Seal (1968) and Kendall (1975) presented case-studies comparing the results of component analyses using a covariance and a correlation matrix. A variety of case-studies is given in Morrison (1967) and Blackith and Reyment (1971).

The introductory remarks about the structure of the covariance matrix should be borne in mind if a correlation matrix is used. Correlations which are not calculated by ordinary product-moment methods may give rise to impossible eigenvalues (Kendall 1975).

Geometry of the principal component transformation

The linear transformation in equation (5.1) has a simple geometrical interpretation (Morrison 1967; Seal 1968; Jöreskog *et al.* 1976). The N observations can be thought of as being distributed in a p-dimensional attribute-space. If the data are multivariate normally distributed, the points can be enclosed in ellipsoidal (or hyperellipsoidal for more than three dimensions) contours of equal probability density, centred on the common mean of zero. Rotation of an axis to the position where it has maximum variance is equivalent to finding an axis of the hyperellipsoid. The eigenvalues and eigenvectors determine the lengths and orientations of the component axes. If two successive eigenvalues are equal, they present a circular cross-section, and the orientation of the axes is not unique. A zero eigenvalue means that the hyperellipsoid can be represented in fewer than p dimensions.

A bivariate example is illustrated in Figure 5.1. Attributes X_1 and X_2 are correlated, and if the observations are represented as deviations from the means, the points are referred to axes x_1 and x_2. For any point P (x_1, x_2), the new co-ordinates are, by elementary co-ordinate geometry:

$$[y_1 \, y_2] = [x_1 \, x_2] \times \begin{bmatrix} \cos \theta & -\sin \theta \\ \sin \theta & \cos \theta \end{bmatrix} \tag{5.8}$$

i.e. in matrix notation $\mathbf{Y} = \mathbf{XA}$. Thus, the algebraic transformation in equation (5.1) represents a rotation of the axes through an angle θ, the angle of rotation being determined by the maximum variance constraint. That the rotation satisfies the algebraic properties already given is shown

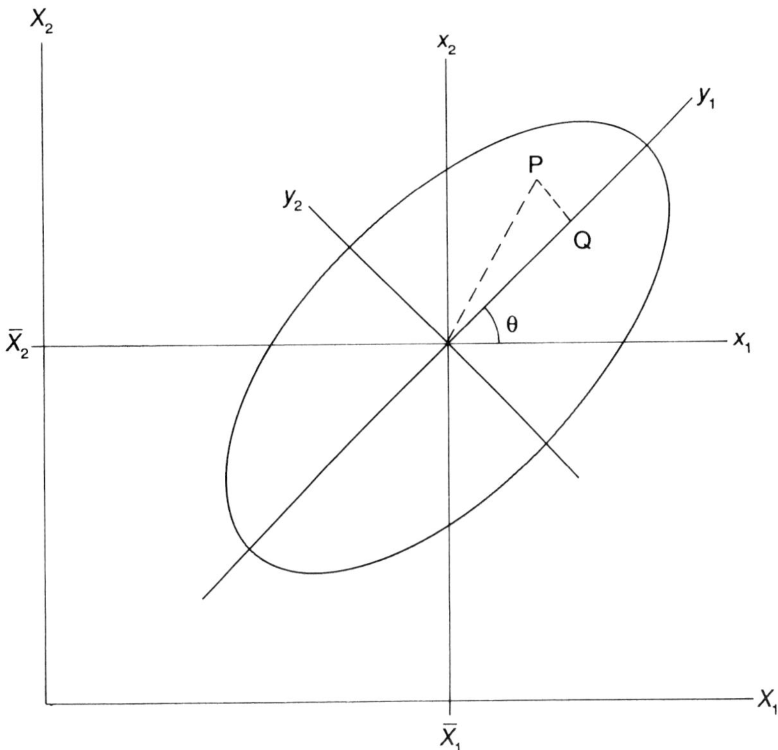

Figure 5.1 The geometry of the principal component transformation

in Morrison (1967 p230). It is easy to verify from (5.8) that

$$\mathbf{AA}' = \mathbf{A}'\mathbf{A} = \mathbf{I}$$

It is clear that in Figure 5.1 the variance of the first component is greater, and that of the second component less, than that of either of the original variates.

The equation for the surface of a hyperellipsoid is

$$\mathbf{a}'\Sigma^{-1}\mathbf{a} = c \qquad (5.9)$$

where c is a constant and $\mathbf{a}$ is a vector of co-ordinates for the points on the surface. The major axis of this hyperellipsoid is defined as the line from the centre to a point on the hyperellipsoid where the squared distance $\mathbf{a}'\mathbf{a}$ is a maximum. To locate this point, we need to maximize $\mathbf{a}'\mathbf{a}$ subject to (5.9). This leads to the result

$$\Sigma\mathbf{a} = \lambda\mathbf{a} \qquad (5.10)$$

which is comparable to equation (5.3). By analogy with that equation, the eigenvectors $\mathbf{a}_j$ or $\mathbf{b}_j$ give the axes of the hyperellipsoid, and the transformations in equations (5.1) and (5.6) represent a rotation of the co-ordinate axes so that the new axes are in the directions of the axes of the hyperellipsoid.

The ellipse for the jth axis is

$$\mathbf{a}_j' \mathbf{D}^{-1} \mathbf{a}_j = \frac{\mathbf{a}_j' \mathbf{a}_j}{d_j} = c \tag{5.11}$$

where c is a constant. The half-length of the jth axis (i.e. the distance from the origin to the point on the ellipse furthest from it) is $\sqrt{(d_j c)}$ (Anderson 1958 p278; Morrison 1967 p82).

It is useful to draw scatterplots showing the positions of objects on pairs of component axes. Such charts are static maps of the positions of objects in component space. A dynamic interpretation would regard the position of each object as being the resultant of pulls due to the vector weightings.

It is not essential to assume that the sample is drawn from a multivariate normal population, unless significant tests are required. Another way of looking at the component axes is that the sum of the squared distances of the points from their projections on each axis is minimized. In Figure 5.1 the projection of point P on to the first axis is Q. For $i = 1$ to N points, the sum of $(P_i Q_i)^2$ is minimized. If we call the origin O, then, by Pythagoras,

$$\sum_{i=1}^{N} (OP_i)^2 = \sum_{i=1}^{N} (P_i Q_i)^2 + \sum_{i=1}^{N} (OQ_i)^2 \tag{5.12}$$

Since the sum of $(P_i Q_i)^2$ is minimized, the sum of $(OQ_i)^2$ is maximized, and this is proportional to the component variance. Since the squared distance between any point and its projection depends on contributions from all dimensions, it is evident that dimensions with small eigenvalues will contribute little to the distance.

Eigenvalues and eigenvectors

We may think of the orthonormal eigenvectors as rectangular axes of co-ordinate systems in which the attributes are vectors at various angles, all passing through the origin. The correlation between any pair of vectors is the cosine of the angle between them, and can be calculated as the inner product of the vectors normalized to unit length. It is necessary to remember that these angles are angles in the complete space of the analysis. If some dimensions are omitted, any study of these angles will require a new matrix of direction cosines (e.g. see Hope 1968 p52). The geometrical, and other, aspects can be illustrated by a simple bivariate example which can be worked through with pencil, paper, and a pocket calculator. The data

Table 5.1 Data matrix, expressed as deviations from the attribute means

	$\mathbf{x}_1$	$\mathbf{x}_2$
	−8	−1
	6	10
	−2	−10
$\mathbf{X}\,(8 \times 2) =$	8	1
	0	3
	−6	−6
	0	−3
	2	6
SSQS	208	292
Variance	29.71	41.71

matrix is given in Table 5.1 (see Hope 1986; Jöreskog *et al.* 1976 p49).

Note that because the mean of a variate is zero for data expressed in this way, the sum of $(\mathbf{X} - \bar{\mathbf{X}})^2$ is simply calculated as the sum of the squared values. The sums of squares and products matrix $\mathbf{X}'\mathbf{X}$ is:

$$\begin{bmatrix} 208 & 144 \\ 144 & 292 \end{bmatrix}$$

The eigenvalues of the sums of squares and products matrix are 400 and 100, and their sum is equal to the total sum of squares. The orthonormal eigenvector matrix $\mathbf{A}$ is

$$\begin{array}{cc} & \begin{array}{ccc} y_1 & y_2 & \text{SSQS} \end{array} \\ \begin{array}{c} x_1 \\ x_2 \end{array} & \begin{bmatrix} 0.6 & -0.8 \\ 0.8 & 0.6 \end{bmatrix} \begin{array}{c} 1 \\ 1 \end{array} \end{array} \quad \begin{bmatrix} \cos\theta & -\sin\theta \\ \sin\theta & \cos\theta \end{bmatrix}$$

$$\text{SSQS} \quad 1 \quad 1$$

so that for any point (object) the first and second component values $\mathbf{Y} = \mathbf{XA}$ are

$$y_1 = x_1 \cos\theta + x_2 \sin\theta \qquad y_2 = -x_1 \sin\theta + x_2 \cos\theta$$

It is easy to verify that: (i) $\theta = 53.13°$; (ii) the component values are as given in Table 5.2; (iii) the sum of the squared component values is equal to the corresponding eigenvalue. The rotation is illustrated in Figure 5.2.

Because the eigenvectors (columns of the matrix) are orthonormal, and the sums of the squared column elements are unity, the square of an element gives the proportion of the sum of squares of a component contributed by an attribute. Thus, attribute x_1 contributes 0.36 or 36% to the first component y_1 and 0.64 or 64% to the second component y_2.

The contribution of a component to an attribute cannot be obtained from the orthonormal eigenvectors, and by concentrating only on the latter the possibility may be overlooked that a particular attribute is accounted

Table 5.2 Component values obtained using the orthonormal eigenvectors from the sums of squares and products matrix

$$\mathbf{Y}\,(8 \times 2) = \begin{bmatrix} -5.6 & 5.8 \\ 11.6 & 1.2 \\ -9.2 & -4.4 \\ 5.6 & -5.8 \\ 2.4 & 1.8 \\ -8.4 & 1.2 \\ -2.4 & -1.8 \\ 6.0 & 2.0 \end{bmatrix}$$

	y_1	y_2
SSQS	400	100

for largely by one of the later components. In order to get more information, other normalizations of the eigenvectors are necessary.

Suppose that we now normalize the eigenvectors so that the sum of squares of each eigenvector is equal to the corresponding eigenvalue. We do this by multiplying each element of an orthonormal eigenvector by the square root of the corresponding eigenvalue as follows:

$$\begin{array}{cc} & \begin{array}{cc} y_1 & \qquad y_2 \end{array} \\ \begin{array}{c} x_1 \\ x_2 \end{array} & \begin{bmatrix} 0.6\sqrt{400} & -0.8\sqrt{100} \\ 0.8\sqrt{400} & 0.6\sqrt{100} \end{bmatrix} \end{array} = \begin{bmatrix} 12 & -8 \\ 16 & 6 \end{bmatrix} \begin{array}{c} \text{SSQS} \\ 208 \\ 292 \end{array}$$

$$\text{SSQS} \quad 400 \quad 100$$

The eigenvector elements are the co-ordinates, on the centralized x axes, of the points where the component axes meet the ellipse. These points are $(12, 16)$ and $(-8, 6)$, and it can be seen from Figure 5.2 that the lengths of the component axes from the origin to these points are equal to the square roots of the corresponding eigenvalues.

In Figure 5.3, the vectors representing attributes x_1 and x_2 have been drawn on the component axes and have lengths equal to the square roots of their original sums of squares. We note that the ends of the vectors are given by the points $(12, -8)$ and $(16, 6)$ because a row vector in the eigenvector matrix above gives the projection of an attribute on the components. This fact can be used in interpretation when making plots of component values, as it shows if collections of objects are associated especially with certain attributes or collections of attributes. We can also verify that, provided that all the columns of the matrix are included, the sum of squares of a row vector of such a matrix is equal to the sum of squares of the corresponding attribute.

If we now divide each element of the previous eigenvector matrix by the square root of the row sum of squares, we get an eigenvector matrix with unit row sum of squares:

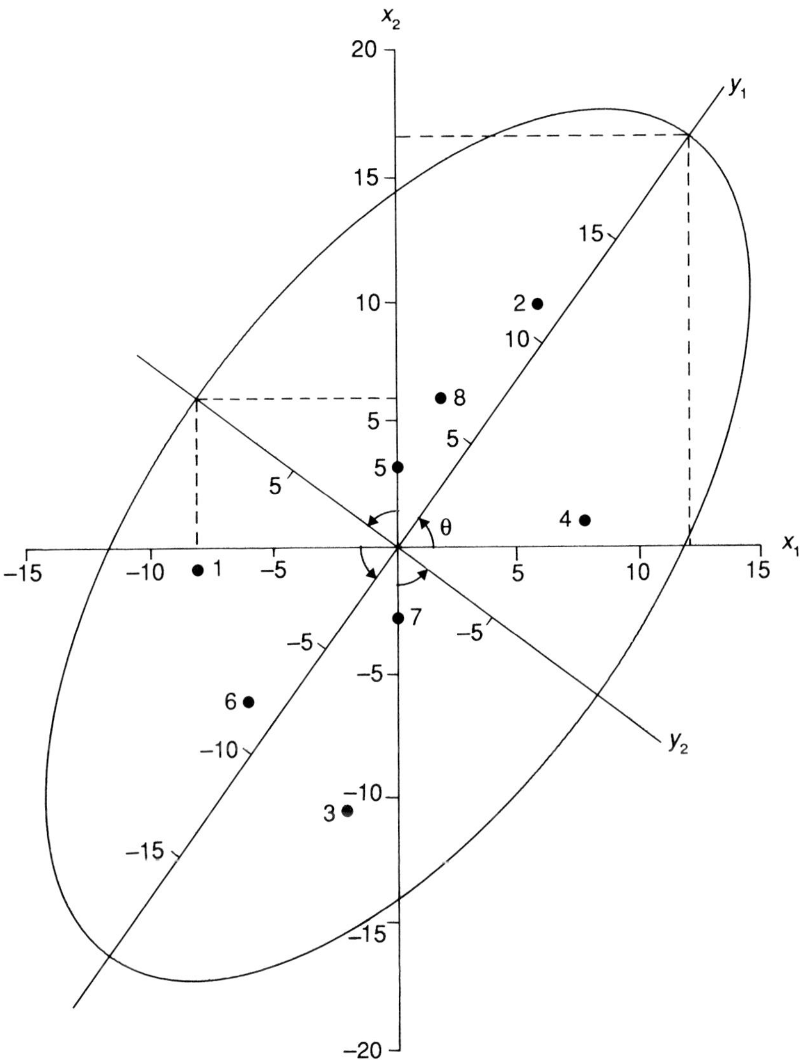

Figure 5.2 Scatterplot showing the positions of points representing the objects in Table 5.1 on axes representing the original attributes (x_1, x_2) and the components (y_1, y_2) obtained from the sums of squares and products matrix of the data in Table 5.1

$$\begin{array}{cc} & \begin{array}{cc} y_1 & y_2 \end{array} \\ \begin{array}{c} x_1 \\ \\ x_2 \end{array} & \left[\begin{array}{cc} \dfrac{12}{\sqrt{208}} & \dfrac{-8}{\sqrt{208}} \\ \dfrac{16}{\sqrt{292}} & \dfrac{6}{\sqrt{292}} \end{array} \right] \end{array} = \begin{array}{c} \left[\begin{array}{cc} 0.8321 & -0.5547 \\ 0.9363 & 0.3511 \end{array} \right] \begin{array}{c} \text{SSQS} \\ 1 \\ 1 \end{array} \\ \begin{array}{ccc} \text{SSQS} & 1.569 & 0.431 \end{array} \end{array}$$

By squaring the appropriate element of this matrix, we obtain the proportion of the sum of squares of an attribute which is accounted for by a given component. This matrix gives the cosines of the angles between the attributes and components, i.e. the direction cosines of the attributes. Remembering that $\cos(90° - \theta) = \sin\theta$, the angles are:

$$
\begin{array}{ccc}
 & y_1 & y_2 \\
x_1 & 33.69° & 56.31° \\
x_2 & 20.56° & 69.44°
\end{array}
$$

Remembering also that the correlation between two vectors is the cosine of the angle made by them at the origin, and can be calculated as the inner product of the two vectors normalized to unit length, it is left to the reader to verify that the angle between the first attribute and the first component is 33.69° (see Figure 5.3). Finally, we can take the eigenvector matrix which has the sums of squares of the eigenvectors equal to the eigenvalues, and divide each element by the corresponding eigenvalue.

$$
\begin{array}{cc}
 & \begin{array}{cc} y_1 & y_2 \end{array} \\
\begin{array}{c} x_1 \\ \\ x_2 \end{array} &
\begin{bmatrix} \dfrac{12}{400} & \dfrac{-8}{100} \\[2ex] \dfrac{16}{400} & \dfrac{6}{100} \end{bmatrix}
\end{array}
=
\begin{bmatrix} 0.03 & -0.08 \\ 0.04 & 0.06 \end{bmatrix}
\begin{array}{c} \text{SSQS} \\ 0.0073 \\ 0.0052 \end{array}
$$

$$
\text{SSQS} \quad \dfrac{1}{400} \quad \dfrac{1}{100}
$$

The sums of squares of the eigenvectors are now equal to the reciprocals of the eigenvalues. If this matrix is used with the data in Table 5.1, the resulting components have unit sums of squares. The reader can check that this is so. The same effect may be obtained by dividing each element in Table 5.2 by the square root of the associated eigenvalue.

We can obtain the covariance matrix by dividing each element of the sums of squares and products matrix by $N - 1$:

$$
\mathbf{S} = \begin{bmatrix} 29.71 & 20.57 \\ 20.57 & 41.71 \end{bmatrix}
$$

Its eigenvalues are $d_1 = 57.137$ and $d_2 = 14.283$, and their sum equals the sum of the variances of x_1 and x_2, i.e. the trace of $\mathbf{S}$. The eigenvalues of a covariance matrix are related to those of the corresponding sums of squares and products matrix by a factor of $N - 1$. The orthonormal eigenvectors are the same as those of the sums of squares and products matrix:

$$
\begin{array}{cccc}
 & y_1 & y_2 & \text{SSQS} \\
x_1 & \begin{bmatrix} 0.6 & -0.8 \\ 0.8 & 0.6 \end{bmatrix} & \begin{array}{c} 1 \\ 1 \end{array} & = \mathbf{A} = \begin{bmatrix} \cos\theta & -\sin\theta \\ \sin\theta & \cos\theta \end{bmatrix} \\
x_2 & & & \\
\text{SSQS} & 1 & 1 &
\end{array}
$$

and they give the direction cosines of the component axes. As the sum of the squared elements of any eigenvector (column) is unity, the square of any element gives the proportion of the variance of that component which is accounted for by the corresponding attribute. However, the contribution of a component to an attribute is not so obvious, for that we need a different normalization.

Because the orthonormal eigenvectors of the covariance matrix are the same as those of the sums of squares and products matrix, the component values in matrix $\mathbf{Y}$ are also the same (Table 5.2) if derived from the mean-centred data of Table 5.1. The difference is that in this case the sum of squares of a component is $N - 1$ times the eigenvalue. If we wanted the components to have sums of squares equal to the associated eigenvalues, we would need to divide each data value in Table 5.1 by $\sqrt{(N - 1)}$. Then, if such a data matrix were premultiplied by its own transpose, the result would be the covariance matrix. It is a useful exercise for the reader to reproduce Figure 5.2 using data scaled in this way, to calculate the component values and their sums of squares, and to show that using that scaling, the type A eigenvector matrix (below) gives the co-ordinates of the ends of the component axes with lengths equal to the square roots of their eigenvalues. A component (column) in matrix $\mathbf{Y}$ could be given unit sums of squares by dividing each component value by $\sqrt{d(N - 1)}$.

If we multiply each element of the orthonormal eigenvector matrix by the square root of the corresponding eigenvalue, the sum of squares of each eigenvector equals the corresponding eigenvalue:

$$
\begin{array}{ccccc}
 & y_1 & y_2 & & \text{SSQS} \\
x_1 & \begin{bmatrix} 0.6\sqrt{57.137} & -0.8\sqrt{14.283} \\ 0.8\sqrt{57.137} & 0.6\sqrt{14.283} \end{bmatrix} & = & \begin{bmatrix} 4.535 & -3.024 \\ 6.047 & 2.268 \end{bmatrix} & \begin{array}{c} 29.71 \\ 41.71 \end{array} \\
x_2 & & & & \\
 & & \text{SSQS} & 57.132 \quad 14.288 &
\end{array}
$$

Provided that all the columns are included in the matrix, the row sums of squares are equal to the variances of the corresponding attributes. For convenience, we can call this a type A eigenvector matrix.

The best-fitting q-dimensional subspace to a scatter of points in a higher (p) dimensional space passes through the centre of gravity of the points, and the sum of squares of the perpendiculars from the points to the subspace is a minimum. Such a space is defined by the centre of gravity (i.e. the mean vector) and the orthonormal eigenvectors. Any point in the subspace can be referred to the p original axes by the relation:

$$\mathbf{x}'_i = \bar{\mathbf{x}}' + \mathbf{k}'_i \mathbf{c}_i \tag{5.13}$$

where $\mathbf{x}'_i$ is a p-dimensional row vector of the raw data matrix $\mathbf{X}$, $\bar{\mathbf{x}}'$ is the mean vector, $\mathbf{k}'_i$ is a row vector of a matrix ($N \times q$) of coefficients, and $\mathbf{c}_i$ is a column vector of

$$\mathbf{C} = \mathbf{AG} \tag{5.14}$$

where $\mathbf{G}$ is a diagonal matrix of positive square roots of the eigenvalues, i.e. matrix $\mathbf{C}$ is a type A eigenvector matrix. The elements of the eigenvectors

$$\mathbf{c}_j = \mathbf{a}_j \sqrt{d_j} \tag{5.15}$$

are 'typical points', summarizing the deviations from the centre of gravity of the original points (Rao 1964 p336). In some types of study, the typical points prove to be useful (e.g. Simonds 1963). In other types of study, they may not admit any useful interpretation. Taylor (1977 p107) plotted these values as 'variance profiles'. Each observed profile is a composite of several such profiles.

However, we cannot use this matrix in the way that we used that obtained in the analysis of the sums of squares and products matrix, i.e. as co-ordinates on the attribute and component axes in Figures 5.2 and 5.3. To do that we need the row sums of squares to equal the attribute sums of squares. If we multiply each element of the type A eigenvector matrix by $\sqrt{(N - 1)}$, we satisfy that condition:

$$
\begin{array}{cc}
& \begin{array}{cc} y_1 & \quad\quad y_2 \end{array} \\
\begin{array}{c} x_1 \\ x_2 \end{array} &
\begin{bmatrix} \sqrt{7} \times 4.535 & \sqrt{7} \times -3.024 \\ \sqrt{7} \times 6.047 & \sqrt{7} \times 2.268 \end{bmatrix}
\end{array}
=
\begin{array}{c}
\begin{bmatrix} 12 & -8 \\ 16 & 6 \end{bmatrix} \\
\begin{array}{cc} \text{SSQS} \quad 400 & 100 \end{array}
\end{array}
\begin{array}{c} \text{SSQS} \\ 208 \\ 292 \end{array}
$$

The sums of the squares of the eigenvectors are now equal to $N - 1$ times the eigenvalues. If, as is usually the case with a covariance matrix, the component values are computed by applying the orthonormal eigenvectors to the mean-centred data matrix, the component values will have sums of squares equal to $N - 1$ times the corresponding eigenvalues. The square root of this value is the length of the eigenvector from the origin to the ellipse (Figure 5.2). For convenience, we can call this a type B eigenvector matrix.

The eigenvector elements give the co-ordinates, on the mean-centred attribute axes, of the points where the ellipses described in equation (5.11) cut the component axes. The row vectors of this matrix give the co-ordinates, on the orthogonal component axes, of the ends of vectors representing the attributes. It can be useful, in interpretation, to mark the

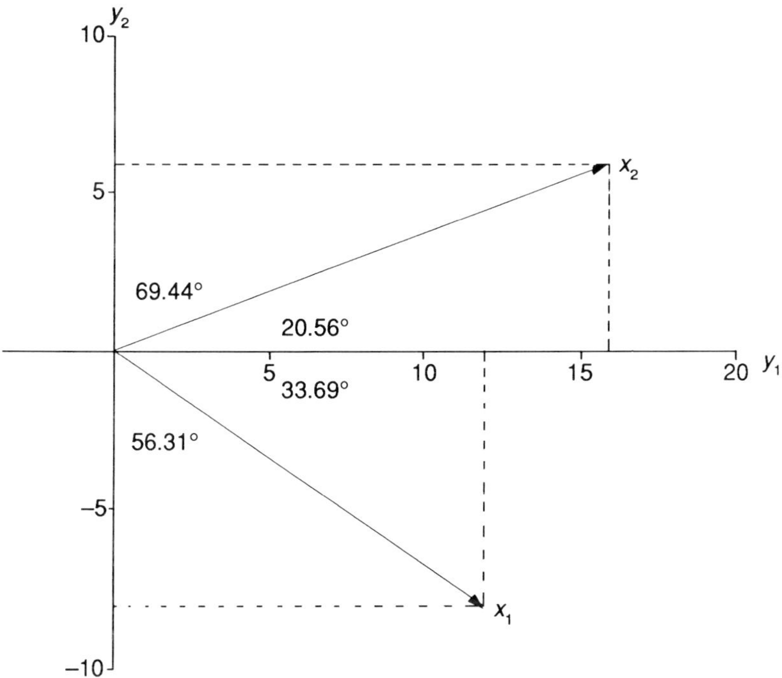

Figure 5.3 Vectors representing attributes x_1 and x_2 drawn with respect to component axes y_1 and y_2 obtained from the sums of squares and products matrix of the data in Table 5.1

ends of the attribute vectors on component plots to find if objects, or collections of objects, are associated especially with certain attributes or groups of attributes.

If we divide each element of a row of a type A matrix by the square root of the variance of the corresponding attribute:

$$\begin{array}{cc} & \begin{array}{cc} y_1 & y_2 \end{array} \\ \begin{array}{c} x_1 \\ \\ x_2 \end{array} & \begin{bmatrix} \dfrac{4.535}{\sqrt{29.71}} & \dfrac{-3.024}{\sqrt{29.71}} \\ \dfrac{6.047}{\sqrt{41.71}} & \dfrac{2.268}{\sqrt{41.71}} \end{bmatrix} \end{array} = \begin{bmatrix} 0.832 & -0.555 \\ 0.936 & 0.351 \end{bmatrix} \begin{array}{c} \text{SSQS} \\ 1 \\ 1 \end{array}$$

$$\text{SSQS} \quad 1.568 \quad 0.431 \qquad 1$$

The rows of this matrix have unit sums of squares, provided that all the columns are included. By squaring the appropriate element, we get the proportion of the total variance of an attribute which is accounted for by a particular component. The elements will be

$$\frac{a_{ij}\sqrt{d_j}}{s_i}$$

55

Each is the correlation coefficient of the ith attribute and the jth component, and a direction cosine of an attribute. For convenience, we can call this a type C eigenvector matrix.

This matrix is the same as that derived in the analysis of the sums of squares and products matrix, because the angle of rotation is the same. Only the eigenvalues, i.e. the lengths of the components, are different.

Finally, for completeness, by analogy with the last normalization of the eigenvectors of the sums of squares and products matrix, we could divide each element of the type A matrix by the corresponding eigenvalue:

$$
\begin{array}{c} x_1 \\[2em] x_2 \end{array}
\begin{array}{cc} y_1 & y_2 \end{array}
\begin{bmatrix} \dfrac{4.535}{57.132} & \dfrac{-3.024}{14.288} \\[1.5em] \dfrac{6.047}{57.132} & \dfrac{2.268}{14.288} \end{bmatrix}
=
\begin{bmatrix} 0.0794 & -0.2117 \\ 0.1058 & 0.1588 \end{bmatrix}
\begin{array}{c} \text{SSQS} \\ 0.0511 \\ 0.0363 \end{array}
$$

$$
\text{SSQS} \qquad \frac{1}{57.137} \qquad \frac{1}{14.283}
$$

This is equivalent to dividing each element of the orthonormal eigenvector matrix by the square root of the corresponding eigenvalue. If this matrix was used, together with the mean-centred data matrix with each value divided by $\sqrt{(N-1)}$, then the resulting components would have unit sums of squares. In practice we normally calculate the component values from the mean-centred data and the orthonormal eigenvectors, and the sum of squares of a component is $N-1$ times the eigenvalue. If we wanted the components to have unit sums of squares, we could also divide each component value by $\sqrt{d(N-1)}$.

We can calculate a correlation matrix as the inner products of the attribute vectors normalized to unit length. The reader can verify that it is:

$$
\mathbf{R} = \begin{bmatrix} 1 & 0.584 \\ 0.584 & 1 \end{bmatrix}
$$

Note that 0.584 is the cosine of 54.25°, which is the angle between attribute vectors x_1 and x_2 (see Figures 5.3 and 5.5). The eigenvalues are $l_1 = 1.584$ and $l_2 = 0.416$. The sum of the eigenvalues is equal to the trace of the correlation matrix. The orthonormal eigenvector matrix $\mathbf{B}$ is:

$$
\begin{array}{c} x_1 \\ x_2 \end{array}
\begin{array}{cc} u_1 & u_2 \end{array}
\begin{bmatrix} 0.7071 & -0.7071 \\ 0.7071 & 0.7071 \end{bmatrix}
\quad \theta = 45°
$$

The rotation is illustrated in Figure 5.4.

If there are only two attributes, the eigenvalues are $(1 + r)$ and $(1 - r)$ where r is the correlation coefficient. The elements of the first eigenvector

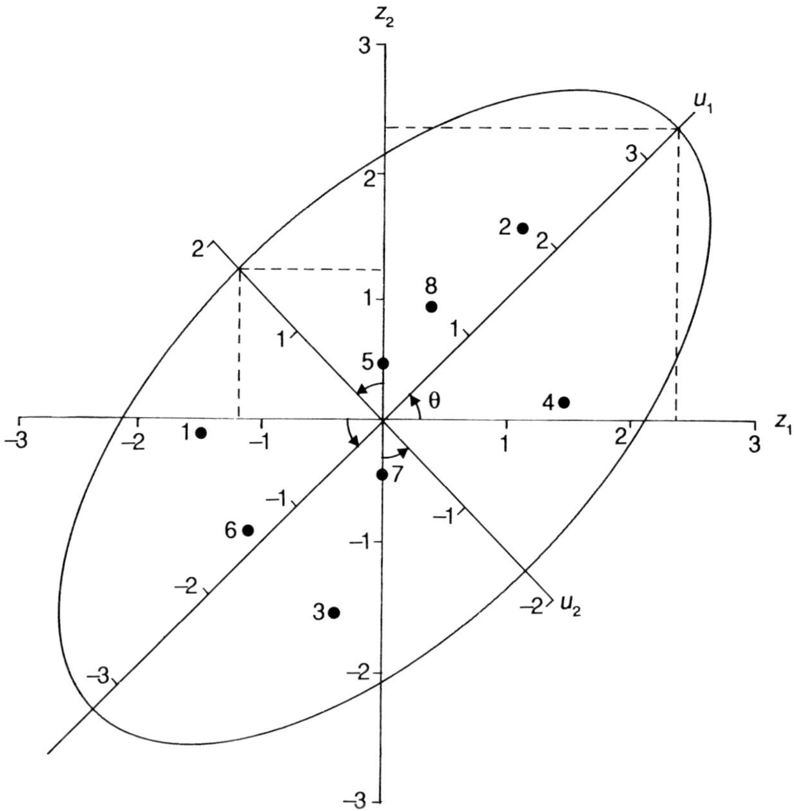

Figure 5.4 Scatterplot showing the positions of points representing the objects in Table 5.3 plotted in relation to axes representing the standardized attributes (z_1, z_2) and the components (u_1, u_2) obtained from the correlation matrix

are both equal to $1/\sqrt{2}$, the elements of the second eigenvector are $1/\sqrt{2}$ and $-1/\sqrt{2}$ or have the signs reversed (Chatfield and Collins 1980 p 63).

The orthonormal eigenvector matrix bears no simple relation to that derived from the covariance matrix, although here, too, the eigenvectors are the direction cosines of the component axes and the square of any element gives the proportion of the variance of that component which is accounted for by the corresponding attribute. Again, the contribution of a component to an attribute needs a different normalization. The variance-standardized data are given in Table 5.3.

The component values are obtained from $\mathbf{U} = \mathbf{ZB}$, the reader can verify that they are as given in Table 5.4. Allowing for small rounding errors, the sum of squares of the component values is $N - 1$ times the corresponding eigenvalue.

Table 5.3 The variance-standardized data

	z_1	z_2
	-1.47	-0.15
	1.10	1.55
	-0.37	-1.55
$\mathbf{Z}(8 \times 2) =$	1.47	0.15
	0	0.46
	-1.10	-0.93
	0	-0.46
	0.37	0.93
SSQS	7.02	7.00
Variance	1	1

Table 5.4 Component values calculated from the variance-standardized data using the orthonormal eigenvectors

	u_1	u_2
	-1.15	0.93
	1.87	0.32
	-1.36	-0.83
$\mathbf{U}(8 \times 2) =$	1.15	-0.93
	0.33	0.33
	-1.44	0.12
	-0.33	-0.33
	0.92	0.40
SSQS	11.13	2.91

By multiplying each eigenvector element by the square root of the corresponding eigenvalue, the eigenvectors can be normalized so that the sum of the squared elements of an eigenvector equals the corresponding eigenvalue, the new eigenvector elements being $b_{ij}\sqrt{l_j}$:

$$
\begin{array}{cc}
u_1 & u_2
\end{array}
$$

$$
\begin{array}{c} z_1 \\ z_2 \end{array}
\begin{bmatrix} 0.7071\sqrt{1.584} & -0.7071\sqrt{0.416} \\ 0.7071\sqrt{1.584} & 0.7071\sqrt{0.416} \end{bmatrix}
=
\begin{bmatrix} 0.8899 & -0.4561 \\ 0.8899 & 0.4561 \end{bmatrix}
\begin{array}{c} 1 \\ 1 \end{array}
$$

$$
\begin{array}{ccc}
\text{SSQS} & 1.584 & 0.416
\end{array}
$$

The sum of squares of each row is unity, and by squaring the appropriate element, we obtain the proportion of the total variance of an attribute which is accounted for by a particular component. Each element of this matrix is the correlation of an attribute with a component, and is a direction cosine of an attribute. For convenience, we can call this a type D eigenvector matrix. In order to reproduce the equivalent of Figure 5.2 for a correlation matrix, we first plot the points on axes z_1 and z_2, i.e. the variance-

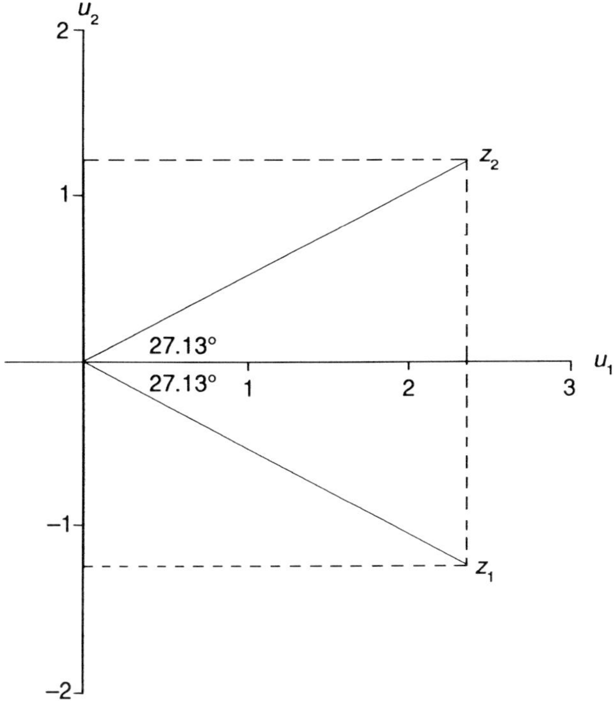

Figure 5.5 Vectors representing the standardized attributes of Table 5.3 drawn with respect to component axes u_1 and u_2 obtained from the correlation matrix

standardized variates. If we multiply the elements of the type D eigenvector matrix by $\sqrt{(N-1)}$ we get

$$
\begin{array}{cccc}
 & u_1 & u_2 & \text{SSQS} \\
z_1 & \begin{bmatrix} 2.354 & -1.207 \\ 2.354 & 1.207 \end{bmatrix} & & \begin{matrix} 7 \\ 7 \end{matrix} \\
z_2 & & & \\
\text{SSQS} & 11.083 & 2.914 &
\end{array}
$$

Within small rounding errors, the sums of squares of the eigenvectors are now equal to those of the component values (Table 5.4). It can be seen in Figure 5.4 that these eigenvector elements are the co-ordinates, on the standardized data axes, of the points where the component axes meet the ellipse. The squared lengths of the component axes from the origin to those points are equal to $N-1$ times the corresponding eigenvalues. For convenience, we can call this a type E eigenvector matrix.

Figure 5.5 is analogous to Figure 5.3, with the important difference that the attribute vectors z_1 and z_2 are of equal length. This is because the data are standardized to have the same variance. In Figure 5.2, the lengths of the attribute vectors x_1 and x_2 are proportional to their variances. Within small rounding errors, the cosine of the angle beween the attribute vectors should in each case be 0.584, which is the correlation coefficient. Furthermore, the rows of the type E eigenvector matrix give the co-ordinates on the component axes which define the ends of the vectors representing the variates (the row sums of squares are equal to the attribute sums of squares). It could be useful in interpretation to draw these vectors on plots of points on component axes to find if objects, or collections of objects, are associated especially with certain attributes or groups of attributes.

It is left to the reader to try other normalizations and their effects, using the same principles as those used for the analysis of the covariance matrix. Those given above are of practical value in interpretation.

In principal component analysis, the eigenvector elements are direction cosines. As the signs of vectors are arbitrary, it is interesting to consider the effects of changing the signs of the elements of an eigenvector. In the following simple example (Figure 5.6a), we have a point P(1, 2) on original axes x_1 and x_2. The eigenvector matrix is

$$
\begin{array}{cc} y_1 & y_2 \end{array}
$$
$$
\begin{array}{c} x_1 \\ x_2 \end{array}
\begin{bmatrix} 0.6 & -0.8 \\ 0.8 & 0.6 \end{bmatrix}
\begin{bmatrix} \cos\theta & -\sin\theta \\ \sin\theta & \cos\theta \end{bmatrix}
$$

This is a rigid rotation, and it is clear that $\theta = 53.13°$. To find the co-ordinates of P on the new axes we have

$$
y_1 = x_1\cos\theta + x_2\sin\theta = (1 \times 0.6) + (2 \times 0.8) = 2.2
$$
$$
y_2 = -x_1\sin\theta + x_2\cos\theta = (1 \times -0.8) + (2 \times 0.6) = 0.4
$$

Consider now the effect of changing the signs in the second eigenvector:

$$
\begin{array}{cc} y_1 & y_2 \end{array}
$$
$$
\begin{array}{c} x_1 \\ x_2 \end{array}
\begin{bmatrix} 0.6 & 0.8 \\ 0.8 & -0.6 \end{bmatrix}
$$

It is evident that nothing has been changed in the first eigenvector, so that the first component is unchanged. However, in the second eigenvector we have $-\sin\theta_2 = 0.8$, so that $\sin\theta_2 = -0.8$, and $\cos\theta_2 = -0.6$. These values correspond to $\theta_2 = -126.87°$ (Figure 5.6b). Thus, we see that changing the eigenvector signs reverses the direction of the corresponding component axis. We can verify that

$$
y_2 = -x_1\sin\theta_2 + x_2\cos\theta_2 = (1 \times 0.8) + (2 \times -0.6) = -0.4
$$

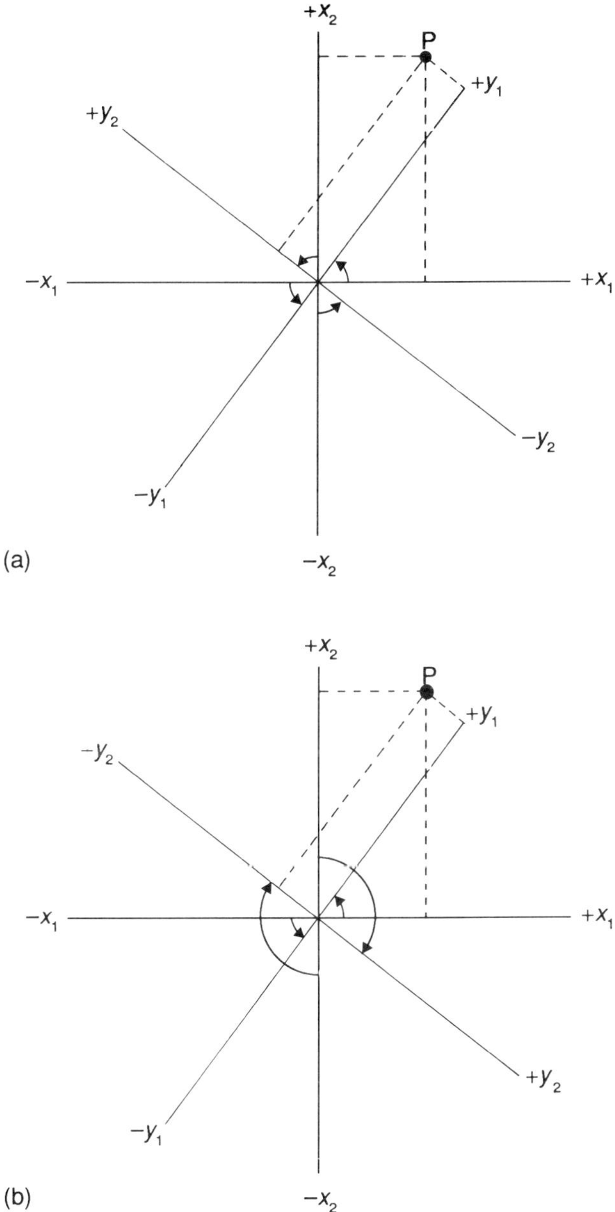

Figure 5.6   Illustrates the effects, on a principal component rotation, of changing the signs in the second eigenvector

We can generalize from two dimensions to q dimensions, but we need some new notation. Considering a rigid rotation such as that in Figure 5.6a.

$$\phi_{11} = \text{angle between } x_1 \text{ and } y_1$$
$$\phi_{12} = \text{angle between } x_1 \text{ and } y_2 \quad (= 90° + \phi_{22})$$
$$\phi_{21} = \text{angle between } x_2 \text{ and } y_1 \quad (= 90° - \phi_{11})$$
$$\phi_{22} = \text{angle between } x_2 \text{ and } y_2 \quad (= \phi_{11})$$

We know that $\sin \phi = \cos (90° - \phi)$ and $-\sin \phi = \cos (90° + \phi)$, therefore

$$\sin \phi_{11} = \cos \phi_{21}$$
$$-\sin \phi_{11} = \cos \phi_{12}$$

So that

$$y_1 = \quad x_1 \cos\theta + x_2 \sin\theta \text{ becomes } x_1 \cos\phi_{11} + x_2 \cos\phi_{21}$$
$$y_2 = -x_1 \sin\theta + x_2 \cos\theta \text{ becomes } x_1 \cos\phi_{12} + x_2 \cos\phi_{22}$$

Extending to q dimensions

$$y_1 = x_1 \cos\phi_{11} + x_2 \cos\phi_{21} \text{------} + x_q \cos\phi_{q1}$$
$$y_2 = x_1 \cos\phi_{12} + x_2 \cos\phi_{22} \text{------} + x_q \cos\phi_{q2}$$
$$y_n = x_1 \cos\phi_{1n} + x_2 \cos\phi_{2n} \text{------} + x_q \cos\phi_{qn}$$

Theory for the fixed case: Singular value decomposition of the data matrix

This is another way of looking at the theory, and it is given in many texts (e.g. Jöreskog *et al.* 1976; Chambers 1977). It illustrates the robustness of principal component analysis as a least-squares approach to producing a 'best' q-dimensional reconstruction of the data matrix. We have already seen (equation 4.1) that the general factor analysis model for the data matrix is

$$\mathbf{X} = \mathbf{FH'} + \mathbf{E}$$

where $\mathbf{F}$ ($N \times q$) is a matrix of factor scores, $\mathbf{H}$ ($p \times q$) is a matrix of factor loadings, and $\mathbf{E}$ ($N \times p$) is a matrix of residuals or error terms which are assumed to be uncorrelated with the factors. Furthermore, in principal component analysis, the residual terms are assumed to be small, so that

$$\mathbf{X} \simeq \mathbf{FH'}$$

(equation 4.4). Applying the method of least squares, we wish to minimize

$$\| \mathbf{X} - \mathbf{FH'} \|^2 \tag{5.16}$$

and we want the $N \times p$ matrix represented by $\mathbf{FH'}$ to be of rank $q < p$. Such a matrix $\mathbf{Q}$ is called a *lower rank least-squares approximation* of $\mathbf{X}$, and is given by

$$\mathbf{Q} = g_1\mathbf{w}_1\mathbf{a}_1' + g_2\mathbf{w}_2\mathbf{a}_2'\ldots\ldots + g_q\mathbf{w}_q\mathbf{a}_q' \qquad (5.17)$$

that is, the first q terms of the singular value decomposition corresponding to the q largest eigenvalues. This does not give a unique solution to $\mathbf{F}$ and $\mathbf{H}$, because of the transformational indeterminacy of factors (see, e.g. Jöreskog *et al.* 1976). However, we are free to choose a solution, and if we let our estimate of $\mathbf{H}$ be $\mathbf{A}$ and of $\mathbf{F}$ be $\mathbf{WG}$ then the covariance matrix of the factors is

$$\mathbf{F'F} = \mathbf{GW'WG} = \mathbf{GG} = \mathbf{D} \qquad (5.18)$$

where $\mathbf{D}$ is a diagonal matrix containing the eigenvalues of $\mathbf{S}$ as in equation (5.2).

Alternatively, we could choose $\mathbf{H} = \mathbf{AG}$ and $\mathbf{F} = \mathbf{W}$. In this case, $\mathbf{F'F} = \mathbf{W'W} = \mathbf{I}$, so that the components will be not only orthogonal, but also standardized to unit length. Also, $\mathbf{H'H} = \mathbf{GA'AG} = \mathbf{GG} = \mathbf{D}$, as in equation (5.14).

A measure of the closeness of the approximation in equation (5.17) is

$$d_{q+1} + d_{q+2}\ldots\ldots + d_r \qquad (5.19)$$

that is, the sum of the $r - q$ smallest eigenvalues. The relative measure

$$\frac{d_{q+1} + d_{q+2}\ldots\ldots + d_r}{d_1 + d_2 + d_3\ldots\ldots + d_r} \qquad (5.20)$$

or some related function, may also be used (Jöreskog *et al.* 1976). Rao (1964 p334) pointed out that while the overall measure might have a small value, the overall configuration of the points might be distorted due to outlying points far away from the best-fitting space, and he suggested that it may be useful to compute the length of the perpendicular of each point on the best-fitting space.

Theory for the random case: Least squares approximation to the covariance matrix

The general factor analysis model for the covariance matrix (equation 4.3) is

$$\Sigma = \mathbf{H}\Phi\mathbf{H}' + \Psi$$

In principal component analysis, Ψ is assumed to be small, and with uncorrelated factors $\Phi = \mathbf{I}$ and we have

$$\Sigma = \mathbf{HH}' \qquad (5.21)$$

A reasonable criterion for fitting the model is to find the best least squares approximation, that is, to minimize

$$\|\mathbf{S} - \mathbf{HH}'\|^2 \qquad (5.22)$$

where **S** is the sample covariance matrix. Matrix **T**, the lower rank least squares approximation to **S**, is given by

$$\mathbf{T} = d_1\mathbf{a}_1\mathbf{a}_1' + d_2\mathbf{a}_2\mathbf{a}_2'\ldots\ldots + d_q\mathbf{a}_q\mathbf{a}_q' \tag{5.23}$$

where d_j are the eigenvalues of **S** and $\mathbf{a}_j$ are the corresponding eigenvectors. In matrix form, equation (5.23) is

$$\mathbf{T} = \mathbf{ADA}' \tag{5.24}$$

which is our estimate of **HH**'. If we let

$$\mathbf{H} = \mathbf{AG} \tag{5.25}$$

where **G** is a diagonal matrix of singular values, i.e. of $\sqrt{d_j}$, then our estimate of **H** is a type A eigenvector matrix (cf. equation 5.14).

Thus, as successive components are extracted from **S**, the matrix **T** can be calculated, and the closeness of the approximation is as equations (5.19) and (5.20). Taylor (1977) suggested the proportionate fit coefficient

$$\frac{1 - \|\mathbf{S} - \mathbf{T}\|^2}{\|\mathbf{S}\|^2} \tag{5.26}$$

This approach emphasizes that it is the covariation of the entire set of variates which is being summarized, and that the optimality is a property of the q-dimensional space spanned by the components, rather than being the property of individual components. The unweighted least squares criterion requires all residuals to be equally weighted, and the residual variances should be the same for all the attributes. This will usually be so if all attributes are measured in the same units and the measurements are of the same order of magnitude (Jöreskog *et al.* 1976).

Choice of attributes to be included in the analysis

In a sense, this is the most important of all the stages, since it corresponds to a hypothesis about the relevance of the various attributes. In the absence of properly-planned multivariate investigations of biological variability, such data as might be available, or which are readily measurable, tend to be used. While useful information may arise from such data, this is no justification for the researcher to shirk the responsibility for formulating proper hypotheses about the attributes which are likely to be important. In choosing the attributes it should be remembered that principal component analysis is not invariant under change of scale, so that the units of measurement are important. An example showing how the results and interpretation of a principal component analysis depend upon the form of the initial data is given by Press (1972 pp296–301).

In outlining the eigenvalue decomposition of a covariance or correlation matrix it was stated that it is preferable for the matrix to be of full rank. We do not wish to complicate the analysis by including, as separate attributes, weighted averages or other linear compounds suggested by earlier studies, since these will reduce the rank of the matrix and obscure whatever latent structure may be present (Morrison 1967 p222).

Type of data

Principal component analysis is a method for partitioning variances. Although it can be applied to sums of squares and products matrices, it is more commonly applied to covariance and correlation matrices. In deciding whether or not this approach to principal component analysis can be used, it is necessary to know if a sensible covariance or correlation matrix can be computed from the data (see Chapter 4). The Pearson product-moment correlation coefficient assumes that the data are, at least approximately, normally distributed. If the data are continuous, or are discrete but can be treated as if they were continuous (section 3.1), but are markedly non-normal, a non-parametric correlation coefficient may be used (e.g. see Howard and Howard 1981). However, care is necessary when applying principal component analysis to correlation matrices which are not calculated by ordinary product-moment methods (Kendall 1975).

Gower (1967b) discussed the special cases of variates of constant sum and ranked data.

Binary (presence–absence) data have a binomial distribution, although by the central limit theorem this will approach a normal distribution if the number of objects is sufficiently large. The effects of binary data on correlation matrices were discussed in Chapter 4, in general they are not suitable for this approach to principal component analysis. One way of avoiding this problem is to convert the data to a multiple contingency table and to use this in calculating correlation coefficients (e.g. Ivimey-Cook and Proctor 1967). Another way of dealing with pure binary data is to calculate the eigenvalues and eigenvectors of $\mathbf{X'X}$ where the data matrix $\mathbf{X}$ may or may not be centred (Noy-Meir 1973). Gower (1966 p332) gave the justification for a principal component analysis of the matrix of corrected sums of squares and products between the variates. He showed that this is equivalent to assuming that the objects are represented by points whose distances apart are proportional to $\sqrt{(1 - S_{ij})}$, where S_{ij} is the simple matching coefficient between objects i and j.

It is important to note that ordered multistate data which are obtained by arbitrarily scoring observations such as absent, rare, common, abundant, as 0, 1, 2, 3, will not contain distance information. That is, although abundant is scored as 3 and rare as 1, the ratio 3:1 does not represent the actual quantities of the objects. Such data are not suitable for principal component analysis, but they can be used in principal co-ordinate analysis.

An extension of principal component analysis to mixed measurement types was described by Young *et al.* (1978). This method has not yet found wide use, but it represents part of a recent trend to develop multivariate methods for the analysis of categorical data (see Nishisato 1980).

Transformation of the data

The choice of whether or not to transform the data is a subjective one. Some investigators are led to the decision to transform by consideration of the theoretical basis of their measurements, others may be guided by the requirements for tests to be applied at a later stage. Generally, it is probably better not to transform the data unless there is a real reason for doing so.

In the analysis of a covariance matrix, prior logarithmic transformation of the data matrix is often used. Logarithms reduce differences between standard deviations, so that effects of scale and magnitude of the original attributes are reduced. Logarithmic transformation has the effect of giving measurements with the same proportional variability the same variance, so that no data standardization is then necessary (Marriott 1974; Pimental 1979). It is also useful in transforming counts (e.g. Cassie 1963). It converts the analysis from one of linear combinations of the attributes to one of their ratios, and so can be particularly useful in growth studies (Gould 1967).

Principal component analysis assumes linearity in the data. The question arises as to what should be done if the data are not linear. The usual answer is to try to linearize the data by some form of transformation. Unfortunately, it may be easy to find a transformation which will linearize one pair of attributes, but not so easy to find one which will linearize them all (Kendall 1975 p28).

Calculation of eigenvalues and eigenvectors

The main computational problem of principal component analysis is to solve the eigenstructure of a symmetric matrix. Since the use of digital computers became widespread, significant advances have been made in the development of fast and accurate programs. Modern methods first reduce the matrix to tridiagonal form by Householder's method. The eigenvalues of the tridiagonal matrix are the same as those of the original matrix, the eigenvectors are not, but are readily transformed back. At the time of writing, the best method for finding all the eigenvalues and eigenvectors of a symmetric matrix is to use the TRED2 routine of Martin *et al.* (1971) for the tridiagonalization and the TQL2 routine of Bowdler *et al.* (1971) to calculate the eigenvalues and eigenvectors. This combination produces eigenvectors which are always accurately orthogonal, even for multiple eigenvalues. The order in which the eigenvalues are found is, to some extent, arbitrary. The accuracy of individual eigenvectors is, of course, dependent on their inherent sensitivity to changes in the original data (Wilkinson 1965).

As a consequence of the singular value decomposition (section 2.3), the eigenvalues, eigenvectors, and component values can be obtained directly from the data matrix, provided it is suitably scaled and centred.

Dimensionality and sphericity

In principal component analysis, with p attributes we normally obtain $r \leqslant p$ non-zero eigenvalues. If $N \leqslant p$, we obtain a maximum of $N - 1$ components. For many purposes, it is convenient to disregard components which have small variances, treating them as constants. Hence, the main components of variation may be studied in a subspace of dimension $q < p$. Although some information is lost in this approach, the q components may account for enough of the original variance to make the reduced dimensionality useful. Indeed, much of the variability in the lower components may only be noise. However, before components are rejected, the possibility should be considered that one of the lower components may be important to a particular attribute or to a particular sub-group of objects. In choosing to ignore a component, we are not doubting its reality. Any dimension which exists in a sample must exist in the population, and however small the eigenvalue, it could not have arisen from a population in which the corresponding value is zero (Kendall 1975). What we are saying is that that part of the variance is not relevant to the particular study in hand. The question of relevance cannot be decided by a mathematical test.

With some data, the first eigenvalue is very large, and successive eigenvalues decrease sharply. It is then clear how many components are likely to be of practical importance. In many cases, however, the result is not so clear. The first component may account for half, or less, of the total variance, and successive eigenvalues decrease gradually. This may be due to one of two causes. Firstly, the variances of the original measurements are approximately equal (or have been made to be). Secondly, the correlations between the original measurements may be strongly curvilinear (Williams 1976).

Morrison (1967) stated that, in his experience, if the first four or five components do not account for something like 75% of the variance, there is usually little point in extracting further vectors, for even if the later eigenvalues were sufficiently distinct to allow easy computation of the components, the interpretation of those components may be difficult or impossible. He suggested that it is frequently better to summarize the data in terms of the first components with large, markedly distinct, variances, and to include as highly specific and unique variates those represented by high loadings in the later components, although the latter are likely to be associated with considerable noise from the other, unrelated, variates. Marriott (1974) suggested that if the attributes have been standardized, it may be reasonable to discard all components with a smaller variance than that of a single attribute, but, on the whole, it is better to retain unimportant components than to discard information of value.

Jeffers (1965) proposed a rule of thumb test based on the fact that if the correlation matrix is an identity matrix, i.e. no attribute correlates with any other attribute, the eigenvalues will all be unity. If there are correlations, some of the eigenvalues will all be unity. If there are correlations, some of the eigenvalues will be greater than unity and some will be less. Hence, a component with an eigenvalue less than unity represents a component which accounts for a smaller proportion of the variance than would be represented by each of the basic attributes separately. In practice it has been found useful to consider only those components having eigenvalues equal to or greater than unity, although we might also look at the next one or two components with eigenvalues less than unity, provided that they are greater than about 0.75. Kaiser argued that a variety of statistical and practical considerations suggest that a rule-of-thumb test based on eigenvalues greater than unity (or on the sum of s_i^2/p for a covariance matrix) is the best single test in any analysis of structure, in either principal component analysis or factor analysis (see also Ibanez 1973). This rule, which seems to have been initiated by Guttman (1954), appears to work well with small or moderate-sized samples (Cooley and Lohnes 1971 p104). Again, we should bear in mind the possibility that the bulk of the variance of a particular attribute or subset of objects may be attributable to a component with a small eigenvalue.

If the first few eigenvalues of a correlation matrix are small, it may be because most of the elements of the correlation matrix are small and principal component analysis is not a suitable technique for those data (Chatfield and Collins 1980 p68). Clearly, if all of the intercorrelations among the attributes are zero, the attributes are effectively components, and in the absence of significant intercorrelations no reduction in dimensionality is possible. Incidentally, components depend on the ratios of the correlations, not on their absolute values. If all the off-diagonal elements of a correlation matrix are divided by a constant, the eigenvalues change but the eigenvectors do not (Chatfield and Collins 1980 p67).

Hawkins *et al.* (1982 p346) stressed the need for care in rejecting components with small eigenvalues if the components are to be used in cluster analysis.

It is a good idea to calculate the proportion of the variance of each attribute by each component before deciding whether or not to reject components with small eigenvalues. This is easily done using a type C or D eigenvector matrix, and it will show whether or not ignoring a particular component means throwing away more information about one attribute than the others (e.g. see Mardia *et al.* 1979 p224; Pimentel 1979 p66). Reyment (1979) discussed the interpretation of components with small eigenvalues, and gave some examples in biology and geology. He concluded that such components can provide useful information, e.g. on the occurrence of invariant growth relationships in organisms or on proportional relations in elements in geological samples.

If redundant attributes are included in a principal component analysis, the large positive correlations between attributes and their redundant counterparts can lead to major components with spuriously-large eigenvalues. The redundant attributes will be detected in the components with small eigenvalues. Also, an outlier along a minor component indicates an observation whose multivariate structure differs from that of the rest of the population, whereas an outlier along the largest principal component differs in overall 'size' (Hawkins and Fatti 1984).

A different question is that of sphericity. The well-known chi-square test of Bartlett, with its various modifications (Kendall 1975; Lawley and Maxwell 1971), tests whether some or all of the eigenvalues are equivalent. Applied to all the eigenvalues, the test has different interpretations according to whether it is applied to a correlation matrix or a covariance matrix. Applied to a correlation matrix, the test asks whether the eigenvalues show any tendency to deviate from a uniform value of unity. If they do not, the correlations among the tests cannot be assumed to differ from zero, and there is no point in transforming the data to orthogonal components. If the test is applied to a covariance matrix, it is a test of the independence of the variates and of the equality of their variances (Hope 1968).

Such statistical tests require a large sample drawn from a p-dimensional multivariate normal population with a covariance matrix having p distinct, non-zero, eigenvalues. The asymptotic distribution theory of the eigenvalues and eigenvectors of correlation matrices is much more complicated than that of covariance matrices (Anderson 1963). Furthermore, it does not follow that all the components which reach statistical significance in a large sample necessarily remove a very large proportion of the variance, and so some of them may be comparatively unimportant in practice (Bartlett 1950 p80). Krishnaiah and Lee (1977) have discussed the problem of the testing of eigenvalues of covariance matrices. James (1977) described tests of the hypothesis that a prescribed subspace is spanned by principal components. The values of such tests in the analysis of real data is doubtful, and many workers prefer to use the rule of thumb approach given above.

Let us consider that the measurements x are composed of systematic or 'true' values x_1 and errors of measurement x_2. If the errors x_2 are random, we can assume that the components of x_2 are uncorrelated and have the same variance, i.e. x_2 has the covariance matrix $\sigma^2 \mathbf{I}$. The components of x_1 may vary in an r-dimensional space and have a covariance matrix Ψ of rank $r < p$. Then $\Sigma = \Psi + \sigma^2 \mathbf{I}$, and the $p - r$ smallest roots of Σ are σ^2. We consider testing the hypothesis that the $p - r$ smallest roots are equal. In this case, it is essential for all the variates to have the same units. In the analysis of a covariance matrix, the null hypothesis is true if the error variances are equal. In the case of a correlation matrix, the null hypothesis is true if the error variances are proportional to the systematic variances (Anderson 1963 p139). Rao (1964 p349) gave an example of the way in which principal component analysis can be used to detect heterogeneity in readings from nominally identical instruments, provided that the instrument errors are independent and have nearly the same distributions.

Jöreskog *et al.* (1976) suggested that a residual correlation matrix may be calculated after each factor or component has been extracted. When the residual matrix consists of correlations solely due to random error, no further components need be extracted. The standard error of the residual correlations, estimated roughly as $1/\sqrt{(N-1)}$, can be used to find the cut-off point at which the standard deviation of the residuals becomes less than the standard error.

One approach which might prove useful, but which does not seem to have received much attention, is the use of non-parametric tests, for example to examine the success of an eigenvalue in differentiating between objects.

Reification and rotation

Reification is the interpretation of the results of a mathematical analysis in terms of the original problem and observations. Sometimes, some or many of the components can be fairly readily interpreted in terms of features of the original observations. Components may correspond to features which have already been appreciated in an intuitive way. It is important to realize the limitations of this type of reification. The order in which the components appear, the proportion of the variance associated with each, and, hence, the conclusions drawn, depend upon the objects chosen and the observations included in the analysis, as well as on sampling variations. The stability of the components of a sample can be examined by making similar sets of measurements on several samples, and doing a separate analysis on each. Kendall (1975 p28) pointed out that the direction cosines are rather sensitive to fluctuations such as one would get from one sample to another, especially where two eigenvalues are close together. He therefore considered it to be unwise to place too much emphasis on the numerical value of any particular coefficient in an eigenvector. Craddock (1973) examined the use of principal component analysis in the analysis of meteorological data, and pointed out that the results of principal component analysis are not stable from one data set to another. This was considered by Cattell (1965) to be a point in favour of factor analysis.

In some types of analysis, the first component has the character of a size vector, while others are shape vectors. Jolicoeur and Mosimann (1960), in studies on the painted turtle, emphasized that for the first component to be a size vector, all coefficients must be of the same sign. Rao (1964) gave a mathematical justification for this conclusion. Examples of other studies of this type are given in Blackith and Reyment (1971), and the topics of shape and growth were discussed by Pimentel (1979). However, the first component should not be regarded as a size component unless the structure of the observations clearly suggests such an interpretation. For example, in studying heath vegetation (species presence–absence) Ivimey-Cook and Proctor (1967) found that the first component reflected species abundance.

A useful aid in the interpretation of component plots is to mark the ends of the attribute vectors on the plots. It is then possible to see if particular objects or groups of objects are associated particularly with certain attributes. The required co-ordinates are obtained from a type B or type E eigenvector matrix. It is also useful to find, on any component, objects which have (absolute) component values greater than the component standard deviation. Such objects most markedly display the attributes associated with the components (Mather 1976 p236). Sometimes, the attribute vectors alone are plotted (e.g. see Legendre and Legendre 1983). However, this is not as useful as superimposing the attribute vectors on a scatterplot of points representing the objects.

The variance of an object is the square of its distance from the origin, i.e. the sum of its squared component values. It is thus possible to calculate the relative contribution of an object to the variance in a component, and to detect if an object is an outlier (see e.g. Pimental 1979 p72).

For detecting lack of fit of individual observations to the q-dimensional principal component analysis model, Rao (1964) suggested computing the sum of the squared lengths of the projections of the observations on to the last $p - q$ component axes, i.e. for the ith object,

$$d_i^2 = \sum_{j=p-q}^{p} (\mathbf{x}_i' \mathbf{a}_j)$$

where $\mathbf{x}_i'$ is the row vector of observations for the ith point (centred or standardized as necessary) and $\mathbf{a}_j$ is the jth column eigenvector. A large value of d^2 would either suggest an aberrant observation or a poor fit. Other possibilities were suggested by Gnanadesikan (1977 p261) and Anderson (1971).

Sometimes, the individual components do not suggest a simple reification, that is, the pattern of eigenvector elements on a particular component is not susceptible to a simple biological or ecological interpretation. Holland (1969) pointed out that while the individual components themselves may be of little biological significance, the space defined by their vectors is, and it is only a matter of geometrical manipulation to determine the extent to which the vectors of other components corresponding to biological hypotheses, or derived from other bodies of data, lie within such a space. Holland further suggested that if certain hypothetical or consistent components can be found to fit the observations, it may be preferable to define the space occupied by the observations in terms of these new components. If this is done, two problems arise. Firstly, the variation associated with each new component will be unknown. Secondly, the number of such components may not be sufficient to define the space. He suggested ways of overcoming these problems (see also Pimentel 1979 p70).

Pimentel (1979) discussed the use of principal component analysis in making comparisons between groups ('multigroup component analysis').

Various methods have been proposed for rotating the component axes to give a new set of axes spanning the same space. The proportion of the total variance accounted for by the first q components will be unchanged, but the variances of the individual new components will be different from the originals, and they will not be subjected to maximum variance constraints, i.e. they will not be the axes of the hyperellipsoid as defined in equation (5.9). Furthermore the rotated axes may not be orthogonal, so that the variances of the new components do not constitute a partition of the total variance attributable to the component space. Hence, great care is needed in interpreting such components. If axes are to be rotated, the aims of the rotation may be clearly stated and formulated in mathematical terms, so that the angle of rotation can be determined. In many studies, researchers apply one or more of the standard rotation techniques used in factor analysis.

For example, Ivimey-Cook and Proctor (1967), studying heath vegetation, found that in principal component analysis of the correlation matrix, the first three components reflected species abundance, soil moisture, and base status respectively. The fourth and fifth components were not readily interpretable. The first five components accounted for nearly 88% of the total variance. Examination of the ordination charts suggested that a rotation of the axes might bring them into line with configurations of points, and so the components were rotated according to Kaiser's varimax criterion to give five new axes which were readily interpretable as corresponding to five recognizable vegetation types. On the other hand, with some data varimax rotation may not bring about an improvement. Mucina and Polacik (1982) found that the first three component axes of vegetation data from a transition mire were ecologically interpretable, but after varimax rotation the interpretability of the first axis was unclear because of the effects of some outlier samples.

Noy-Meir (1970) found little advantage in oblique rotation, but Carleton (1980) thought it useful to study the results of both orthogonal and oblique rotation. Harris (1975 p 191) discussed the effect of rotation on the component values.

Another method of rotation involves the fact that in an association of organisms which is virtually a pure stand, the entropy of mixing of the constituent elements is low relative to that of an association where many elements are approximately equally represented (Pelto 1954). Zones of intergradation between two associations or communities have high entropy (Howarth and Murray 1969), and if components are rotated to positions such that the entropy of the system is minimized, the resulting new axes might be expected to link the centres of associations (McCammon 1966, 1968). Pelto's method is suitable for percentage data.

Swain *et al.* (1979) devised a synthetic problem in order to discover if 'factor analysis' methods (in this case principal component analysis) are capable of finding a 'correct' set of underlying factors. The variables used were seven properties of right circular cylinders, all of which were determined precisely by only two underlying factors, the radii and heights of the cylinders. To be successful, the method must separate these factors, and must not suggest any others. Swain *et al.* found that principal component analysis correctly identified that there were two dimensions, but some of the eigenvector weightings did not make any physical sense. Varimax rotation did not produce any improvement, apparently because the concept of simple structure was not applicable to this particular problem. However, a more complex rotation procedure devised by Swain *et al.*, and involving subsidiary problem-related information, gave the correct eigenvector weightings. Harris (1975) also gave an example which showed that there is no guarantee that principal component analysis will uncover the true generating processes which gave rise to the observed variates.

The use of components in regression analysis

A serious problem in regression analysis is collinearity, or approximate collinearity. The estimators of the coefficients depend upon the inverse of the covariance matrix of the regressors. If one variate is a linear function of another, the coefficients in a regression equation which includes them are indeterminate, for the determinant of the covariance matrix vanishes. If some of the eigenvalues of the covariance matrix are small, its determinant is also small, and the coefficients will be ill-determined (Kendall 1975). Stepwise multiple regression of a system which has high internal correlations is statistically dubious, there is no certainty that the result will be optimal.

The use of orthogonal component values in multiple regression not only simplifies computations. Interpretation of the results is facilitated by the partitioning of the overall R^2 into uncorrelated sources of predictability. However, as Harris (1975) pointed out, this gain in interpretability will be illusory if the components do not lend themselves to simple substantive interpretations.

Massy (1965) discussed principal component regression analysis in some detail. In an empirical study, Massy found that the principal component regression technique gave larger values for R^2 while employing fewer regressors, than did their classical counterparts in three out of four cases examined. He concluded that the methods involved are useful because (i) they permit rapid calculation of the correlations between a dependent attribute and each of the components, and (ii) they refer the regression results back to the projections of the original independent attributes in the space spanned by the components included in a given regression. This partially overcomes the problem of identifying the components in order to give meaning to the regression coefficients.

Daling and Tamura (1970) used components not as new attributes, but as the reference frame to identify a near orthogonal subset of explanatory attributes. Selection of such attributes minimizes overlapping of information supplied by explanatory attributes in the regression. They suggested applying varimax rotation to a type D eigenvector matrix. The varimax criterion produces a matrix of vectors in each of which a few attributes tend to have high loadings while the rest have small or zero loadings. The dependent attribute is then regressed on each varimax factor to identify the explanatory attributes which, having minimum interdependence among themselves, appear to make a significant contribution to the variance of the dependant attribute. Hawkins (1973) described an interactive method which has the advantage of enabling alternative good subsets of predictors to be found easily and is useful for establishing the amount of leeway the user has for forming subregressions. In this method, the elements of the orthonormal eigenvectors of a correlation matrix are divided by the square roots of the corresponding eigenvalues, and the new eigenvector matrix is subjected to a varimax rotation, the rotated matrix being used to suggest possible attributes (see also Hawkins and Fatti 1984).

Page and Fabian (1978) used principal component regression to examine the relationships between disease and air pollution. While the use of components in this way may be less 'efficient' than is canonical correlation, this approach may be easier to interpret, and can thus yield more information. In particular, it is possible to see if groups which can be recognised in one battery of data can be related to features of another.

The use of components in cluster analysis

In some problems, we may wish to examine whether the N objects can be classified into groups, or clusters, so that the points within a cluster are close together, but the clusters themselves are, ideally, far apart. The problems involved in cluster analysis are discussed in later chapters, but here we can discuss briefly how principal component analysis can be of use. The discriminatory power of principal components can serve as a clustering technique of great generality (examples in Blackith and Reyment 1971). Plotting points on pairs of orthogonal component axes can help in several ways. Firstly, it may suggest the suitability (or otherwise) of a particular form of analysis, for example if there are clearly-defined and separated groups and whether these are spherical or elongated. Secondly, it may show why a particular technique has not given satisfactory results, and it may suggest alternatives. Finally, it may confirm that a suggested clustering looks reasonable and fits the observations realistically.

The orthogonality of the principal component transformation means that it is distance- and angle-preserving in an r-dimensional space. How well plots of points on pairs of orthogonal component axes represent the real configuration of the points depends on how well this configuration is preserved in the reduced space. One way of looking for distortions in such plots is to superimpose on them the minimum spanning tree (Chapter 15).

It can also be useful to plot histograms of the frequency distributions of points along selected components. This will show whether or not there is multimodality along a component (e.g. see Webster and Burrough 1972a Fig.6).

In the analysis of a covariance matrix, if most of the total variance is accounted for by the first two components, Gabriel's (1971) biplot can be useful in showing interobject distances and in indicating clustering of objects, as well as displaying variances and correlations of the variables. Ter Braak (1983) discussed biplots in relation to plant species diversity.

The investigator may have decided to accept the first q dimensions as preserving sufficient of the total variance for practical purposes. The configuration of the points in this q-dimensional space can be studied by calculating distances between pairs of points. It might be worth also computing the distances in r-dimensional space, and to examine the differences between the distances in the two spaces to see if they are uniformly small (Rao 1964).

Chang (1983) showed that for data representing a large sample from a mixture of two multivariate normal distributions with a common covariance matrix, the information content of each component is not a monotonic function of its eigenvalue, and the component with the largest information content does not necessarily have the largest eigenvalue. In terms of differentiating between groups, i.e. in using components for cluster analysis, the best subset of q components is that with the q largest values of Δ_{si}, which is the Mahalanobis' distance between the two populations on the ith component. Δ_{si} is a monotonic function of $(\mathbf{p}_i'd)^2/\lambda_i$, where $\mathbf{p}_i$ is the ith eigenvector, d is the difference between the population means, and λ_i is the ith eigenvalue. With a large sample, one can perform likelihood ratio tests for $\Delta_{si} = 0$ against $\Delta_{si} > 0$, and the q components with the largest likelihood ratios can be selected. Real data are unlikely to be multivariate normal with common covariance matrices, so these tests are unlikely to be generally useful in practice. Nevertheless, this does highlight the need to examine pairwise scatterplots of the component values before selecting components for cluster analysis. Some components with large eigenvalues may not be helpful in discrimination, while one or more components with small (and possibly rejected) eigenvalues may be useful.

Since the components are orthogonal, the distances used in clustering can be simple Pythagorean distances. As Pythagorean distance depends on the scale of the attributes, it is unlikely to have much meaning if some attributes have a much greater range of values than others. Hence, it is generally used only when all the measurements have been standardized in some way (Marriott 1974). The components are dimensionless, but as calculated using the orthogonal eigenvectors they have different variances, proportional to the eigenvalues. Hence, if component values are to be used for the calculation of Pythagorean distances, it may be preferable to normalize the components to unit sums of squares (Hope 1968 p56). In effect, this is a question of the weighting given to each component in its contribution to the Pythagorean distance. By normalizing each component, we are effectively weighting it according to its variance.

Rohlf (1970) pointed out that principal component analysis will give reliable representation of intergroup distances over large distances (i.e. between clusters) but will not be reliable over small distances (i.e. among closely-spaced objects within clusters). As the opposite seems to be true of most forms of cluster analysis, principal component analysis is a useful complement to them.

Other uses of components

The orthogonality is useful in other subsequent analyses such as multivariate analysis of variance (Manova). In Manova, the orthogonality may simplify computation, and may facilitate interpretation by the partitioning of the differences among the groups into their differences on each of the components, provided that the components themselves are interpretable. The reduction in dimensionality can simplify computations in Manova, Hotellings T^2, and canonical variate analysis, all of which require the inversion of the covariance matrix of one or more batteries of variates. If there is linear dependence within a battery of variates, the covariance matrix will be singular and cannot be inverted. Rather than deleting one or more of the variates a more elegant approach would be first to perform a principal component analysis on each battery of variates, then to perform the required analysis on the components. The resulting measure of relationship and test of significance will be identical whether principal component analysis or deletion of variates is used. The decison which approach to use therefore depends on questions of computational convenience or interpretability of the resulting regression weights, discriminant function, or canonical coefficients (Harris 1975).

If there is any reason to hypothesize that an object may belong to one of two or more groups which can be discriminated by a particular component, then the fit between such groups and the component values can be tested by analysis of variance (Hope 1968 p58). If the component values have been calculated using the orthonormal eigenvectors of a covariance or correlation matrix as described above, the total sum of squares of a component is $N - 1$ times its eigenvalue. The between-groups sum of squares can be calculated by squaring the mean component value of each group, multiplying the result by the number in the group, and then summing over all the groups. The within-groups sum of squares can be obtained by subtraction. For exploratory purposes, the more complicated designs in the analysis of variance may prove useful. If the hypothesis is that the groups may be discriminated by reference to two or more components, the multivariate analysis of variance may be useful (Rao 1964 p356), bearing in mind the need to satisfy certain requirements. It is worth bearing in mind the fact that, in practice, it may turn out that some of the components with smaller eigenvalues are better discriminators between groups with respect to the differences between means than are the first few components with the larger eigenvalues. Moore (1965) provided biological examples of the use of component values in analysis of variance.

Other uses are possible if the data can be referred to some sort of geographical grid, for example, in stratified sampling (e.g. Patterson *et al.* 1978).

A problem which often occurs in multivariate studies is how to reduce the number of original attributes which need to be measured. Jolliffe (1972, 1973) discussed eight methods, four of which use principal component analysis. He applied five of the eight techniques to real data, including a multiple linear regression analysis. Mansfield *et al.* (1977) presented a method for reducing the number of interdependent attributes. The procedure first deletes components associated with small eigenvalues of $\mathbf{X'X}$ and then incorporates an analogue of the backward elimination procedure to eliminate the independent attributes, this deletion being based on minimal increases in residual sums of squares.

A procedure for selecting a subset of q attributes which come closest to measuring everything measured by all p attributes was described by Cureton and d'Agostino (1983). This consists of rotating orthogonally so that the first rotated component has a non-zero weighting on only the attribute accounting for the largest proportion of the total variance, then rotating so that the second rotated component has non-zero weightings on only this first attribute and the attribute having the largest residual variance, and so on. For this purpose, a series of Gram-Schmidt transformations is used, the first of order q, the second of order $q - 1$, and the last of order 2. The qth attribute is then the one having the largest $(q - 1)$th residual variance.

Advantages and disadvantages of principal component analysis

The main advantage of principal component analysis is in the robustness of the least squares approach to approximating the data matrix or covariance matrix. It is therefore not important for the data to be multivariate normal unless significance testing is required. Other advantages lie in the relative simplicity of the technique. For example, it is easy to see the contributions made by the attributes to each component. Other advantages are (1) the orthogonality of the components, and (2) the reduction of dimensionality. These are useful in subsequent analyses as discussed above.

The main disadvantage lies in the assumption that any relationships among the original attributes are essentially linear, or at least that any non-linear contribution is small. Problems occur if the data are markedly non-linear, for example along an environmental gradient. The orthogonality of the components implies functional independence only if the objects are normally distributed. Norris (1971) used a simple example with artificial data to show that if one attribute has a linear response to an environmental gradient while a second has a sinusoidal response, the resulting component plot shows the environmental gradient to be curved. Situations in which some of the attributes do not increase or decrease linearly along an environmental gradient proposed as a reification should be examined for functional relationships between components (cf. e.g. Norris and Barkham 1970 p609).

Plant species are thought of as being distributed along environmental gradients with frequency curves which are sigmoid if the peak value is at one end of the gradient, or bell-shaped, but not necessarily symmetrical, if the gradient encompasses the whole range of a species. If a species is responding simultaneously to a number of different factors which vary together along an environmental gradient, the combined response will also be bell-shaped, or at least unimodal. Such general expectations of species response have often been confirmed where it has been possible to measure quantity, density, cover, basal area, or biomass along a known environmental gradient (Whittaker 1967; Gauch and Whittaker 1972a). If only a small part of the gradient is examined, the relationships may be approximately linear for a number of species.

When vegetation data are subjected to ordination, it is usually assumed (or hoped) that the ordination axes will correspond with complexes of environmental factors. However, the bell-shaped species response curves mean that the relationship of species quantity (however measured) to a single environmental complex will contribute to more than one of the component axes, i.e. the second axis does not represent a different set of environmental factors from the first. Furthermore, if the peaks for the different species occur at different points along an environmental gradient, each may contribute to a different component axis, so that the dependence on a single environmental complex may be obscured. Other problems occurring with the use of principal component analysis of vegetation data were discussed by Johnson and Goodall (1979).

However, Werger *et al.* (1983) noted that there are at least three reasons for being somewhat cautious in accepting the Gaussian model of species response curves: (i) Although there are examples of apparently bell-shaped response curves of some species along obvious vegetation gradients, it has been found that many other species in the same vegetation show a variety of more complex responses; (ii) Experimental evidence has shown that the ecological response curves of species are modified in a complex way by varying intensities of competition; (iii) It has been pointed out (Austin 1980) that at least three types of environmental gradient must be recognized, and species responses to these different types of gradient may differ. Knowledge about ecological responses of the component species of populations is essential for understanding the ecological basis of plant community boundaries. Using a clustering technique and principal component analysis on data from a transect in grassland vegetation in the Netherlands, they showed the presence of two types of vegetation boundary, one strong, the other weak and gradual. Furthermore, less than one-third to one-half of the species had Gaussian distributions.

It is thus not surprising that principal component analysis has had a mixed reception in vegetation analysis. Although many authors have dealt with its disadvantages, it has been used successfully in a number of studies (e.g. van der Maarel 1969; Londo 1971; Werger 1978; Werger *et al.* 1978). Problems due to non-linearity must obviously be taken into account, but Nichols (1977) concluded that principal component analysis provides an effective technique for data reduction, and that most of its defects are only of major importance when it is used for indirect gradient analysis. Van der Maarel (1979) commented that principal component analysis could be recommended as a lucid, efficient summarisation method which has proved its use in phytosociology. Allen and Shugart (1983) found that principal component analysis of data from a model of a forest stand, obtained by simulated succession, revealed differences which provided insight into the structure, and proved valuable in uncovering previously-unsuspected ordering principles in the model. New hypotheses generated in this way could be validated simply by returning to the succession model. They pointed out that ordination of field data does not allow the simple validation of new hypotheses in the same way as the return to a clearly-defined model.

While the linearity constraint of principal component analysis has received much attention, there has been less concern with the problems raised by the use of mixed data, and by the additive derivation of component values. Some workers think that multiplicative models might be more appropriate. Various problems and approaches to the application of principal component analysis were discussed by Dale (1975).

For some purposes, the orthogonality of the components is an advantage, and these have been discussed above in connection with uses of components. However, there is no reason why we should assume that the underlying factors for which we are searching should be orthogonal. In nature, many factors act together with sufficient consistency for scientific laws to be formulated. One solution to this may be rotation of components (q.v.), an alternative might be to try the factor analysis model, which is discussed in the following chapter.

A practical example of principal component analysis

Table 5.5 gives values for eleven variables measured on surface (0–20 cm) samples collected from six soils belonging to each of eight soil groups (base-deficient brown earths, base-rich brown earths, brown podzolic soils, podzols, peaty podzols, hill peats, peaty gleys, and gleys) in various parts of the UK (Dr A.F. Harrison, personal communication). The lower half-matrix correlation coefficients are given in Table 5.6. In interpreting correlation coefficients, it is necessary to remember that the statistical significance of a correlation coefficient gives the probability that it is not zero. The magnitude of a correlation coefficient and its importance cannot be assessed statistically. However, the square of a correlation coefficient gives the proportion of the variance which is accounted for by the linear relationship of two variables, hence a coefficient of 0.707 accounts for about half of the variance. In Table 5.6 there are few correlations of that size or greater, although many are very highly significant. Total nitrogen and carbon are strongly correlated ($r = 0.967$, $r^2 = 0.93$), because the carbon is organic carbon and most of the nitrogen is organic. Total nitrogen is also strongly correlated with extractable magnesium. Sand content is strongly negatively correlated with total nitrogen, carbon, and extractable magnesium.

The eigenvalues of the correlation matrix are given in Table 5.7. Only three are greater than unity, together they account for 79% of the total variance, while the first four eigenvalues account for 87%. The orthonormal eigenvectors for the first four components are given in Table 5.8. The first component is essentially a contrast between sand on the one hand and

Table 5.5 Data for eleven variables measured on 48 UK soils (0–20 cm) belonging to eight soil groups

pH	Clay%	Silt%	Sand%	Total % C	N	P	Extractable (mg/100 g) Ca	Mg	K	P
4.8	16.0	23.6	54.7	0.99	0.27	0.067	53	7.1	27.0	0.37
4.3	6.4	23.7	61.1	2.1	0.36	0.066	84	9.5	18.0	1.2
5.0	12.7	20.9	57.4	2.0	0.35	0.13	140	9.4	13.0	1.6
5.4	15.0	17.9	61.1	2.8	0.19	0.069	150	14.0	26.0	4.2
5.65	2.7	12.7	75.4	3.8	0.38	0.072	170	7.0	13.0	0.51
6.5	9.9	25.3	55.0	4.6	0.34	0.069	270	16.0	25.0	0.13
7.7	8.7	8.7	79.6	0.74	0.097	0.041	200	5.4	4.6	1.5
5.9	10.4	18.0	66.2	1.9	0.18	0.046	210	6.1	2.5	0.15
4.9	8.6	11.6	75.9	1.3	0.14	0.061	76	2.0	4.3	1.7
5.6	14.9	28.9	49.3	1.9	0.25	0.058	160	10.0	7.1	0.44
5.4	10.4	16.9	66.8	1.8	0.24	0.06	150	10.0	9.7	0.7
7.2	16.1	17.0	61.5	1.7	0.18	0.021	220	8.3	7.1	0.34
4.6	3.3	14.8	63.9	7.1	0.62	0.1	46	9.9	10.0	0.17
4.8	5.2	14.8	67.0	5.8	0.33	0.052	39	15.0	20.0	0.085
4.2	5.1	17.0	62.9	7.4	0.44	0.076	11	3.9	30.0	0.32
4.2	11.6	28.2	43.2	7.4	0.47	0.07	10	4.2	14.0	0.15
4.4	5.0	16.8	62.2	6.8	0.53	0.064	68	11.0	10.0	0.45
4.9	3.4	15.3	66.3	8.2	0.67	0.053	220	12.0	16.0	0.38
4.5	1.9	2.9	92.6	1.1	0.06	0.007	23	1.7	4.7	0.35
3.9	3.8	6.6	84.1	2.8	0.097	0.01	13	1.6	2.3	0.49
3.25	0.9	6.4	84.1	5.1	0.14	0.008	14	3.8	4.2	0.63
3.1	1.8	1.8	85.4	5.6	0.28	0.007	20	5.1	3.0	0.88
3.2	2.8	6.6	85.3	2.5	0.13	0.011	7	0.82	3.4	0.63
3.1	1.9	3.8	88.6	2.8	0.084	0.015	17	1.3	3.7	0.56
4.8	7.1	13.6	38.3	24.0	0.83	0.064	36	21.0	17.0	0.25
3.8	0.2	2.0	12.8	46.0	1.62	0.08	97	14.0	51.0	4.9
3.0	5.6	7.7	29.7	32.0	1.07	0.06	11	4.9	13.0	1.2
3.4	12.0	13.2	37.8	24.0	0.72	0.043	24	5.2	9.5	0.58
3.8	20.7	33.8	14.5	15.0	0.99	0.063	11	7.2	18.0	0.35
3.2	5.0	4.2	6.8	47.0	1.77	0.081	25	19.0	23.0	3.6
4.4	0.2	0.4	4.4	49.0	2.04	0.048	85	110.0	26.0	5.7
4.0	0.2	0.3	2.5	48.0	1.69	0.058	78	110.0	25.0	7.2
3.8	0.3	0.4	7.3	48.0	1.92	0.083	53	42.0	28.0	4.9
5.0	1.0	2.7	10.3	46.0	1.93	0.081	120	88.0	35.0	0.55
4.1	0.4	0.5	4.1	49.0	1.84	0.064	110	75.0	54.0	4.6
4.3	0.3	1.1	6.6	57.0	1.77	0.049	370	100.0	12.0	0.43
4.3	6.0	18.1	61.9	7.2	0.43	0.048	110	13.0	7.8	0.1
3.5	20.5	12.6	45.9	11.0	0.58	0.047	27	7.2	11.0	0.99
4.6	17.2	20.5	44.3	7.3	0.45	0.088	250	26.0	8.7	0.14
4.0	17.9	13.8	37.3	19.0	0.58	0.055	61	7.8	6.5	0.15
4.1	6.9	6.9	72.2	7.7	0.41	0.051	23	7.2	16.0	0.32
4.9	9.7	13.8	57.5	8.8	0.44	0.054	130	11.0	6.6	0.87
6.6	21.8	21.8	47.4	2.8	0.33	0.22	530	24.0	56.0	12.0
7.1	19.8	24.1	42.1	4.8	0.71	0.15	630	56.0	32.0	3.1
6.5	8.8	31.7	47.5	4.5	0.41	0.08	440	16.0	25.0	2.2
6.0	38.3	28.5	22.2	3.5	0.34	0.095	530	93.0	14.0	3.00
5.9	23.5	17.6	42.9	5.6	0.59	0.066	1100	40.0	26.0	0.57
4.9	36.5	24.9	27.6	3.4	0.32	0.051	350	24.0	29.0	0.53

Table 5.6 Coefficients of correlation between the variables in Table 5.5

	pH	Clay%	Silt%	Sand%	Total %			Extractable (mg/100 g)			
					C	N	P	Ca	Mg	K	P
pH	1.000										
Clay	0.411**	1.000									
Silt	0.485***	0.696***	1.000								
Sand	0.108	− 0.121	0.086	1.000							
C	− 0.385**	− 0.417**	− 0.601***	− 0.816***	1.000						
N	− 0.295*	− 0.350*	− 0.486***	− 0.863***	0.967***	1.000					
P	0.406**	0.342*	0.384**	− 0.347*	0.040	0.169	1.000				
Ca	0.627***	0.540***	0.356*	− 0.159	− 0.152	− 0.059	0.433**	1.000			
Mg	0.094	− 0.013	− 0.271	− 0.707***	0.654***	0.696***	0.213	0.326*	1.000		
K	0.153	0.068	− 0.004	− 0.582***	0.431**	0.521***	0.613***	0.299*	0.412**	1.000	
P	0.119	0.011	− 0.189	− 0.434**	0.378**	0.407**	0.578***	0.203	0.445**	0.666***	1.000

*$p \leqslant 0.05$ (0.285)
**$p \leqslant 0.01$ (0.368)
***$p \leqslant 0.001$ (0.461)

Table 5.7 Eigenvalues of the correlation matrix of the eleven soil variables

Component	Eigenvalue	Percentage of trace	
		Component	Cumulative
1	4.3940	39.95	39.95
2	3.2935	29.94	69.89
3	1.0191	9.26	79.15
4	0.8553	7.78	86.93
5	0.4634	4.21	91.14
6	0.3481	3.16	94.30
7	0.2712	2.47	96.77
8	0.1945	1.77	98.54
9	0.1379	1.25	99.79
10	0.0199	0.18	99.97
11	0.0030	0.03	100.00

Table 5.8 Orthonormal eigenvectors corresponding to the four largest eigenvalues of the correlation matrix of the eleven soil variables

Variable		Eigenvector			
		1	2	3	4
1	pH	0.081	0.416*	−0.014	0.495*
2	Clay	0.108	0.417*	−0.336	−0.331
3	Silt	0.210	0.391*	−0.143	−0.439*
4	Sand	0.416*	−0.093	0.317	0.321
5	Total C	−0.436*	−0.184	−0.144	−0.044
6	Total N	−0.450*	−0.111	−0.143	−0.073
7	Total P	−0.167	0.403*	0.381	−0.198
8	Ext† Ca	−0.043	0.430*	−0.255	0.416*
9	Ext Mg	−0.373*	0.087	−0.360	0.324
10	Ext K	−0.330	0.236	0.346	−0.122
11	Ext P	−0.309	0.184	0.515*	0.106

*Absolute value greater than or equal to 0.75 times the largest absolute value
†Ext = extractable

total C, N, and Mg on the other. Those four variables together account for 70% of the variance in that component. The second component is dominated by extractable Ca, clay, pH, total P, and silt which together account for 85% of the variance in that component. The third component is dominated by extractable P, which accounts for some 26% of the variance in that component. The fourth component is essentially a contrast between pH and extractable Ca on the one hand, and silt on the other.

In Table 5.9 the eigenvectors have been normalized to have sums of squares equal to the corresponding eigenvalues. The row sums of squares are unity for the full matrix, and the square of any element gives the proportion of the corresponding variable which is accounted for by the corresponding component. Total C and N are very largely accounted for by the first three components, and adding a fourth makes little difference.

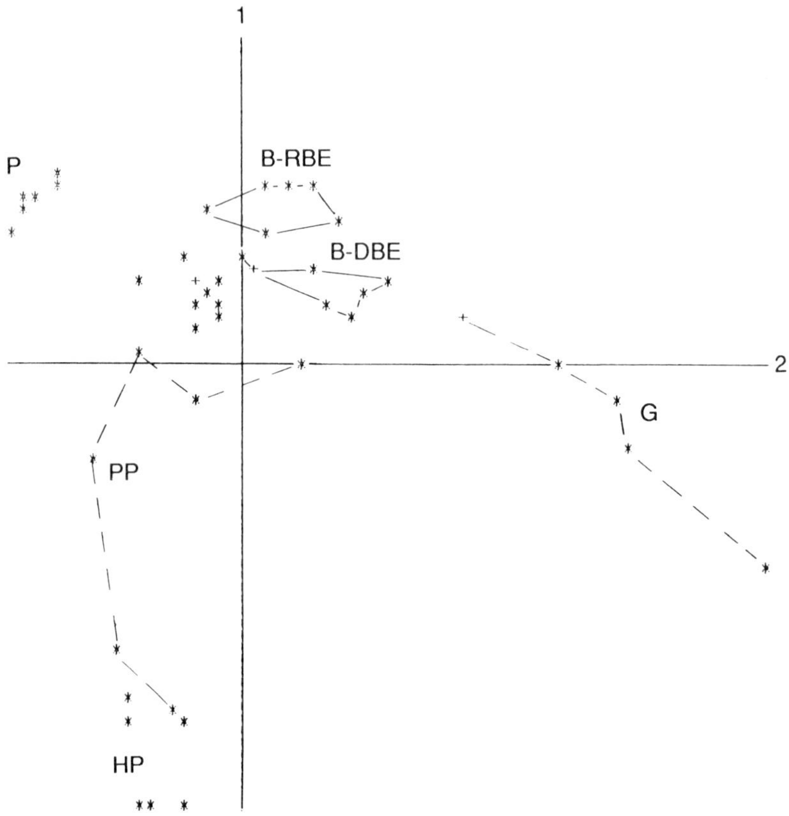

Figure 5.7 Scatterplot of points representing the soils of Table 5.5 on the first two component axes. B-RBE base-rich brown earths, B-DBE base-deficient brown earths, P podzols, PP peaty podzols, HP hill peats, G gleys

Table 5.9 The eigenvectors of Table 5.8 normalized to have sums of squares equal to the corresponding eigenvalues

Variable	Eigenvector for component				Sum of squares	
	1	2	3	4	3 axes	4 axes
pH	0.170	0.755	−0.014	0.458	0.599	0.809
Clay	0.227	0.757	−0.339	−0.306	0.740	0.834
Silt	0.440	0.710	−0.144	−0.406	0.719	0.884
Sand	0.872	−0.169	0.320	0.297	0.890	0.979
Total C	−0.914	−0.334	−0.146	−0.041	0.967	0.969
Total N	−0.943	−0.202	−0.145	−0.068	0.951	0.956
Total P	−0.349	0.731	0.384	−0.183	0.805	0.838
Ext† Ca	−0.089	0.780	−0.258	0.384	0.683	0.831
Ext Mg	−0.781	0.158	−0.363	0.299	0.767	0.857
Ext K	−0.691	0.429	0.349	−0.113	0.784	0.796
Ext P	−0.647	0.334	0.520	0.098	0.801	0.810

†Ext = extractable

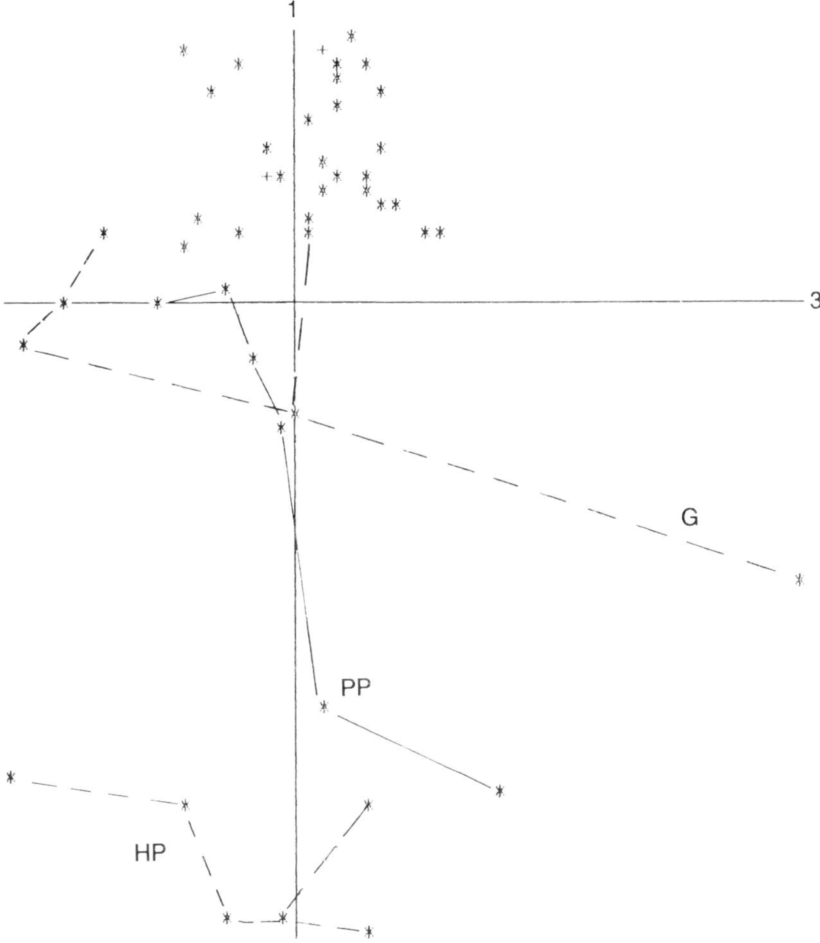

Figure 5.8 Scatterplot of points representing the soils of Table 5.5 on the first and third component axes. PP peaty podzols, HP hill peats, G gleys

Four components account for 80% or more of all the variables.

Points representing the soils are plotted in the space of the first two components in Figure 5.7. The first axis orders the points representing soils which have low carbon (and hence organic matter) and low nitrogen contents at the top of the scatterplot and those with large carbon and nitrogen contents at the bottom. The hill peats, which form a group at the bottom of the first axis, have outstandingly large extractable magnesium contents. The base-rich brown earths, base-deficient brown earths, and podzols form distinct groups, and the brown podzolic soils and peaty gleys form a compact group. However, the peaty podzols and the gleys form extended groups in the first two dimensions.

On the second component axis, the gleys have the largest extractable calcium contents and also have the largest second component values. However, the positions of the points on this axis depend also on the pH and clay, silt, and total phosphorus contents of the soils.

The different soil types are less clearly differentiated on the plot of the first and third component axes (Figure 5.8). In particular, the gleys are strung out along the whole axis, and the soils with positive values on the first axis are less clearly differentiated by the third axis than they are by the second axis.

The component values will be used as an example of hierarchical cluster analysis in Chapter 13.

CHAPTER 6

FACTOR ANALYSIS

The basic principle of factor analysis is that the variance of a variate can be partitioned into the common variance, or communality, and the residual, or unique variance. The latter can be divided into the specific variance and the error variance. The communality is involved in covariances, and communalities are large for variates with large intercorrelations. Examination of the correlation matrix may show that a few of the attributes are highly correlated, while the remainder are not significantly correlated with them or with each other. If the purpose is to seek a basic pattern which summarizes the behaviour of a set of attributes, we might consider retaining those attributes which show high intercorrelations and discarding the remainder.

If we replace the values of unity in the principal diagonal of the correlation matrix with the communalities (i.e. the common factor variances), then some of the eigenvalues, those associated chiefly with the unique variances, would become zero or nearly so. The search for communalities is a major problem in factor analysis. The correlation of an attribute with itself represents an upper bound. As the communality of an attribute is the most that it has in common with other attributes, the square of the multiple correlation coefficient of an attribute with all other attributes constitutes a lower limit (Cattell 1965). The problem is further complicated by the fact that the values of the communalities and the number of factors taken out are inter-dependent, and so cannot be solved simultaneously.

In the past, the procedure adopted was as follows: First, the researcher decided in advance how many eigenvalues and eigenvectors he/she was interested in examining. This in itself presented a problem, there was no way to be sure, in advance, that the correct number had been chosen and, as Williams (1976) commented, it seemed to be a matter of luck. The next step was to form a reduced correlation matrix by replacing the unities in the principal diagonal with estimates of the communalities, such as the squares of the multiple correlation coefficients. Then, the chosen number

87

of eigenvalues and eigenvectors of the reduced matrix were calculated, and the eigenvectors were normalized so that the sum of squares of each eigenvector equalled the corresponding eigenvalue. These eigenvectors provided an improved estimate of the communalities, and the process was iterated until the communality estimates were reasonably constant. This method is no longer used, more efficient methods being now available.

Once the eigenvalues and eigenvectors have been obtained, the calculation of the factor values for the objects is more complex than is the calculation of component values. Also, the structure is commonly further simplified by orthogonal or oblique rotation in the search for 'simple structure'. Factor analysis is computationally more demanding than is principal component analysis, and the former contains a greater element of subjectivity, since certain important decisions have to be taken by the user. Commenting on the value of factor analysis, Mather (1976) pointed out that in recent years considerable theoretical and computational developments have taken place, leading to a new approach with a more solid theoretical basis. Nevertheless, the method should be approached with caution and a clear understanding of what is involved.

Theory

The general factor analysis model (equation 4.1) is

$$\mathbf{X} = \mathbf{FH'} + \mathbf{E}$$

In order to define the model, the following assumptions are necessary: (i) The expected value of each column vector of $\mathbf{F}$ is zero, as is that of each column vector of $\mathbf{E}$; (ii) The column vectors of $\mathbf{F}$ are uncorrelated with those of $\mathbf{E}$; (iii) The residuals, i.e. the elements e_{ij}, are uncorrelated among themselves; (iv) The residual covariance matrix Ψ is diagonal and non-singular; (v) Σ, the covariance matrix of the attributes, has rank $r < p$; (vi) Each attribute is correlated with at least one other.

A basic problem is that neither $\mathbf{H}$ nor $\mathbf{F}$ is uniquely defined for a given data matrix. That is, it is impossible to fix the position of the rotated axes without external criteria. Jöreskog (1963) and others (e.g. Anderson and Rubin 1956) have shown that if equation (4.1) correctly represents the structure of $\mathbf{X}$, then

$$\Sigma = \mathbf{H\Phi H'} + \Psi$$

as in equation (4.3). If we require the factors to be orthogonal, then $\Phi = \mathbf{I}$ and we have

$$\Sigma = \mathbf{HH'} + \Psi \tag{6.1}$$

Unlike principal component analysis, Ψ is not assumed to be small. In factor analysis, Ψ is a diagonal matrix of unique or residual variances. Using the least-squares approach, the problem is to minimize

$$\| \mathbf{S} - \mathbf{HH}' - \Psi \|^2 \tag{6.2}$$

(cf equation 5.22 for principal component analysis). This is equivalent to minimizing the trace of $(\mathbf{S} - \mathbf{HH}' - \Psi)^2$ because Ψ is diagonal.

If we knew Ψ then, by analogy with the theory for the random case given for principal component analysis, the problem of finding the factor loading, or factor pattern, matrix $\mathbf{H}$ would reduce to solving the eigenstructure of $\mathbf{S} - \Psi$. The columns of $\mathbf{H}$ are the eigenvectors normalized so that the sum of squares of each eigenvector equals the corresponding eigenvalue. Unlike principal component analysis, in factor analysis the eigenvectors do not need to be orthogonal. In practice, Ψ is not known and has to be estimated from the data. This is equivalent to choosing the communalities, as the diagonal elements of $\mathbf{S} - \Psi$ are estimates of the communalities (Jöreskog *et al.* 1976).

Various methods were tried in the past (e.g. see Cattell 1965), but now there are new and more efficient methods, although the computational procedures are rather complex.

The unweighted least-squares (ULS) method minimizes

$$U = \text{trace } (\mathbf{S} - \Sigma)^2 \tag{6.3}$$

The ULS solution is equivalent to the traditional iterated principal factor solution and to Harman's Minres solution, which minimizes only the sum of squares of all the off-diagonal elements of $\mathbf{S} - \mathbf{HH}'$ and therefore avoids the communalities. This method is described in Harman (1977) and Mather (1976), and the latter gave a FORTRAN program for its implementation. The ULS approach does not yield a scale-free solution so, in practice, the correlation matrix is used instead of the covariance matrix. The least-squares method does not require the assumption of multivariate normality of the attributes, and statistical tests are not applicable.

The generalized least-squares (GLS) and maximum-likelihood (ML) methods are scale-free. The GLS method minimizes

$$G = \text{trace } (\mathbf{I} - \mathbf{S}^{-1}\Sigma)^2 \tag{6.4}$$

The ML method minimizes the function

$$M = \text{trace } (\Sigma^{-1}\mathbf{S}) - \log|\Sigma^{-1}\mathbf{S}| - p \tag{6.5}$$

As with ULS, we should write $\mathbf{HH}' + \Psi$ for Σ. In the GLS and ML methods, close adherence to the multivariate normal distribution is assumed, although the methods seem to be fairly robust to departures from multivariate normality. One advantage of the GLS and ML methods is that a statistical measure is available for assessing the goodness-of-fit of the model.

A disadvantage is that the computational procedure is rather complicated (see Jöreskog *et al.* 1976; Mather 1976).

Jöreskog (1977) showed that the three functions $U/2$, $G/2$ and M may be minimized by basically the same algorithm. When there is more than one common factor, it is necessary to remove an element of indeterminacy which exists in the basic model before minimizing the function. This indeterminacy arises from the fact that nonsingular linear transformations of the common factors exist which change $\mathbf{H}$ and, in general, Φ, but leave Σ unaltered. Therefore, some additional constraints are necessary in order to obtain a unique set of estimates. In exploratory factor analysis, this is usually done by setting $\Phi = \mathbf{I}$ and either (i) in ULS setting $\mathbf{H'H}$ to be diagonal, or (ii) in GLS and ML, setting $\mathbf{H'\Psi^{-1}H}$ to be diagonal. This leads to an arbitrary set of factors which may then be subjected to a rotation or a linear transformation to another set of factors.

Principal factor analysis

This is probably the simplest form of factor analysis, and was one of the earliest. In it, the first factor is chosen so as to account for as much as possible of the communal variance, the second factor to account for as much as possible of the remaining communal variance, and so on. The method requires suitable estimates of the communalities (for example, squared multiple correlations). It is clear that if the communalities are chosen to be unity, the method reduces to principal component analysis.

Alpha factor analysis

This is a 'non-statistical' method in that the N objects comprise the population, but the p attributes represent a selection from a universe of possible attributes. The aim of this method is to estimate, from the given attributes, common factors which have maximum correlation with the 'universe' common factors. It takes its name from the alpha coefficient of reliability or generalizability (Cronbach 1951), which can be treated in this context as the squared correlation coefficient between a common factor in the selected variates and the corresponding 'universe' common factor. Universe factors associated with negative alpha coefficients are not considered to be meaningful, and the number of factors having positive alpha coefficients is the same as the number of eigenvalues of the correlation matrix which are greater than unity.

A useful feature of this method is that the alpha factor loadings are scale-independent, but a problem is that the communalities generated can exceed unity. This is generally thought to be unacceptable, although Kaiser and Caffrey (1965 p9) did not agree. Alpha factor analysis was described in detail by Mather (1976), who also gave a FORTRAN program.

The use of alpha factor analysis is indicated when the factorial structure of a defined universe of possible attributes is being investigated, and the attributes actually used are a (usually fairly large) random sample. The results do not generalize outside the population of objects studied.

Image analysis

Guttman (1953) introduced the method of image analysis. He suggested that the main difference between it and the common-factor approach is that the latter could be considered to involve the extraction of factors one at a time, and the replacement of the correlation matrix with the matrix of partial correlations of the attributes with the effects of those factors removed. The procedure terminates when the remaining partial correlations are negligible. In image analysis, the multiple correlations between the attributes and the factors are considered rather than the separate partial correlations (see also Harris 1962). The method, which is scale-free, was discussed in detail by Mather (1976), who also gave a FORTRAN program. Both the objects and the attributes are regarded as fixed finite groups, and the results do not generalize.

Choice of attributes, type of data, transformation

Because, in factor analysis, interest centres mainly on the common factors, it does not make sense to include attributes which cannot possibly have some part of their variance in common with others. It is also an important objective to find the simplest factor pattern which can account for the intercorrelations among the attributes. Therefore, the number of common factors should be small and they should be uncorrelated. Each common factor should have as many zero loadings as possible, but if a factor has non-zero loadings on only two attributes, the loadings are not well-determined and this situation should be avoided if possible. The attributes should be chosen so that no two of them appear to resemble each other more closely than they resemble at least one other attribute (Cureton and d'Agostino 1983 p7).

Nabholz and Richardson (1975) emphasized the importance of the choice of attributes in ecological applications, and noted that factor analysis will reflect cruelly an experimenter's inability to choose attributes from a homogeneous collection. They also discussed the problem of sample size, the greater the number of attributes the larger the sample size required. They suggested as a minimum sample size $N = (10p) + 30$, where p is the number of attributes. With small sample sizes, the random errors of the less-reliable correlation coefficients increase the absolute sizes of the correlations in the matrix, resulting in greater communalities and a larger amount of common-factor variance because of the introduction of spurious common-factor variance. Sample size also affects factor loadings and the number of extractable factors. Reliable factor loadings result from improvements in the reliability of the sample correlations, which in turn are improved by increasing the sample size. However, improvement will occur with diminishing returns (Comrey 1973). Hair *et al.* (1979 pp219–234) suggested a minimum of four to five times as many objects as attributes. Perhaps a better guide to a suitable minimum value for N is the chi-squared test given in Seal (1968 p178).

It should also be noted that the unique variance in each attribute is that part which is uncorrelated with all the other attributes. If, to an existing battery of attributes, one or more new ones are added, part of the unique variance in one of the original attributes may become common variance which it shares with one or more of the new attributes.

Factor analysis is concerned with correlations among attributes. The Pearson product-moment correlation coefficient is a valid measure only if the joint distribution of the variates is substantially bivariate normal. Mild leptokurtosis and mild to moderate platykurtosis make little difference. However more extreme departures from normality cause the product-moment correlation coefficient to be an inadequate measure. In factor analysis, normal-standardizing (or ranking as a first approximation) is recommended for variates with substantially-skewed distributions (see Chapter 4). Logarithmic transformation has been used (e.g. Väätänen 1980).

Factor analysis assumes that the relationships between attributes are linear and the input data are on at least an interval scale. Special methods have been prepared for nominal or ordinal data (Green and Tull 1975 p526), but they do not appear to be widely used.

Selecting the number of factors required

The problems involved in deciding how many factors are required to provide an adequate representation of the data are essentially the same as those discussed for principal component analysis. One advantage of the ML and GLS methods, as compared to ULS and other methods, is that there is a chi-square test for assessing the goodness-of-fit of the factor analysis model. A smaller value of chi-square than that tabulated indicates that the number of factors should be reduced, a larger value suggests that it should be increased. This test is illustrated on a practical example in Jöreskog *et al.* (1976 p83). Mather (1976) discussed this test, and noted that it is reliable only if $N - p$ is greater than 50. The test is, in fact, a measure of the degree to which the residual correlations differ from zero.

In Minres factor analysis, the sum of squares of the residual correlations may be printed (in the FORTRAN program of Mather 1976 it is printed as the function value at minimum), but there is no statistical test for this value. In this, and the other forms of non-statistical factor analysis, some rule-of-thumb is necessary. As in principal component analysis, this may take the form of the number of eigenvalues exceeding unity, although this can lead to overfactoring if N is small. Some other suggestions are given in Linn (1968), Jöreskog *et al.* (1976 p126), and Cureton and d'Agostino (1983). Underfactoring, i.e. choosing too few factors, is bad because one or more determinable common factors will be omitted. Overfactoring is bad because factors may be obtained which appear to be interpretable but are in reality due to fortuitous correlations among the attributes. In the maximum-likelihood and Minres methods, if the initial choice of the number of factors turns out to be incorrect, the procedure must be repeated with a new value, as the factors obtained with a value of q are not the same as the first q factors obtained with an initial value greater than q.

Sokal (1959) applied five tests for completeness of factor extraction to four correlation matrices. None of the methods was considered to be entirely satisfactory because when the tests had shown complete extraction significant partial correlations were found in the residual matrices. A test for significance of individual residuals was found consistently to give better results than did the other tests. Everett (1983) showed that split-half factor comparabilities, based on factor scores, can provide a direct and unambiguous method of determining the number of reliable factors that should be retained, and of assessing the appropriate rotation that should be used. The method was considered to be particularly appropriate where the factor scores are to be used as summary or classificatory measures.

If the factors are to be rotated, it is probably better to be generous in the number of factors selected before rotation, as severe distortion will occur if too few factors are chosen (Jöreskog *et al.* 1976). After rotation, unwanted factors can be discarded. The suggested procedure is therefore:

(i) Using one of the computational procedures discussed above, extract as many factors as are required to account for a minimum of 95% to 99% of the total variance, thus obtaining matrix **H**.

(ii) Rotate matrix **H** by an orthogonal rotation procedure, such as varimax (see below).

(iii) Using one of the rules-of-thumb, test the last factor for inclusion. If it does not meet the selected criterion, delete the last unrotated factor, i.e. remove the last column vector from **H**, and go to step (ii). Repeat as necessary, until the matrix has no trivial factors but has not suffered serious distortion due to the removal of too many factors.

Rotation of factor axes

Once a set of q orthogonal factors has been found that accounts for the intercorrelations of the attributes, those factors may be transformed (rotated) to any other set of q factors, not necessarily orthogonal, which account equally well for the correlations. This represents a great indeterminacy in the model, and so some principle is needed on which to base a rotation. Broadly, models for rotation can be based on the effect they produce on either (i) the distribution of the loadings for the attributes, or (ii) the distribution of the factor scores of objects. The former is more widely used, and is discussed here. An example of the latter is criterion rotation, which can be used in a variety of ways (e.g. see Cattell and Khanna 1977). For example, one could rotate until the first factor gives maximum separation of certain groups of objects. In canonical factor analysis, one could rotate to yield maximum prediction from one set of axes to another.

The stability of factors can be tested by performing two factor analyses using measurements of the same attributes on different sets of objects. The assumption is that for the same set of attributes, the same underlying factors will be operating, although probably with different variances in the two samples. Then, if we can find a rotation for each sample such that a factor in one has loadings which are proportional to those of a factor in the other the assumption would be justified. This is known as 'confactor' rotation (see Cattell and Khanna 1977). In practice, it can only be used when there is reason to believe that the real factors are approximately orthogonal.

The positions of the unrotated factors depend to a large extent on the set of attributes used. In order to facilitate comparisons between factor analyses, and to find factors that may be related to important underlying influences, it is desirable to find 'invariant' factors. Rotated factors whose positions are determined by clusters of attributes give a good approximation to factorial invariance (Kaiser 1958). In rotating factors to change the distributions of the attribute loadings some sort of principle is needed upon which to base the rotation. Thurstone (1947) suggested that only those factors for which the attributes have a very simple representation are meaningful, and that an attribute should not depend on all common factors but only on a few of them. That is, the factor loadings matrix should have as many zero elements as possible. Furthermore, a given factor should only be connected with a small proportion of the attributes. This is Thurstone's concept of 'simple structure'. Communalities and unique variances are not affected by linear transformations of factors.

To achieve simple structure, the factors are rotated according to the following criteria: (i) Each row of the rotated matrix should have at least one zero entry; (ii) For q factors, each column of the rotated matrix should have at least q zero entries; (iii) For every pair of columns of the rotated matrix, there should be some attributes which have a zero entry in one of the columns but not in the other; (4) For $q \geqslant 4$, a large proportion of the corresponding elements of every pair of columns should be zero; (5) For every pair of columns, there should be only a small number of attributes with non-zero entries in each column.

Cattell (1952) suggested that if factors are regarded as determinants, then this agrees with Thurstone's principle that each factor should have many attributes with zero loadings, but the former does not require that each attribute should arise from relatively few factors. Furthermore, it also allows the factors to be oblique, whereas Thurstone's principle might call for orthogonal factors (Cattell and Khanna 1977).

The principle of simple structure may be valid for a variety of environmental problems in which, given a set of attributes, it is important to isolate those subsets of attributes which react similarly to the same causal influence. It should go without saying that the search for simple structure must be 'blind'. If the investigator can see, in the plots of the points, a configuration which he/she likes, there is a danger that he/she will rotate to that solution. The alternative to rotation in terms of the quite general hypothesis that the data contain simple structure positions is to rotate directly to test some specific hypothesis. The 'target matrix' represents a hypothesis, and a measure of the goodness of fit of the rotated matrix to the target matrix may help the investigator to decide if the hypothesis is tenable (Mather 1976).

It should be noted that if insufficient factors have been extracted, or if unities were used instead of communalities in the diagonal of the matrix, then the real simple structure will not be obtainable (e.g. see Cattell and Khanna 1977).

Various practical methods have been proposed for finding simple structure (see Crawford and Ferguson 1970 and Cattell and Khanna 1977 for a comparative discussion of the methods). Two of the methods for orthogonal rotation are quartimax, which aims to maximize the fourth power of the loadings, simplifying by rows (attributes), and varimax, which involves simplification by columns (the reason for using the fourth power is illustrated in Cattell and Khanna 1977). The varimax criterion (Kaiser 1958) has become probably the most widely-used rotation procedure. In 'raw' varimax, the variance of the squared factor loadings is used to avoid complications due to signs, thus, for the jth factor:

$$s_j^2 = \frac{1}{p} \sum_{i=1}^{p} (b_{ij}^2)^2 - \frac{1}{p^2} \left[\sum_{i=1}^{p} b_{ij}^2 \right]^2$$

where **B** is the matrix being rotated. Simple structure is achieved when the sum of the individual factor variances is at a maximum, so that the raw varimax criterion is:

$$V_R = \sum_{j=1}^{q} s_j^2$$

Kaiser (1958) suggested that, before the variance is calculated, each row of the matrix should be normalized to unit length by dividing each element by the square root of the associated communality, the communality being equal to the sum of the squared elements of the appropriate row of the factor matrix. After rotation, the rows are restored to their original lengths. This normalization, which has the effect of giving the attributes equal weighting, results in the normal varimax criterion, and it is often found that this gives better results than does the raw varimax criterion (Jöreskog *et al.* 1976; Mather 1976). The normal varimax transformation distributes the total variance more evenly over the rotated factors than does the raw varimax, avoiding the development of an almost-general factor and generally giving a fairly good orthogonal approximation to simple structure (Cureton and d'Agostino 1983).

It has been argued by various researchers that it is unlikely that, in nature, the underlying causal influences which we are seeking will be completely independent. Cattell (1965) pointed out that if orthogonal factors are calculated when, in nature, the underlying influences are really correlated, then the elements of the factor loadings matrix become difficult to interpret. He stated 'in factor analysis, we must distinguish between the independence of a pair of concepts in our minds and their statistical independence in nature'.

If we use oblique factors, the correlations between them can be examined and other, more fundamental, factors can be found. Oblique rotation does not seem to have been much used in ecology, but it has been applied in other earth sciences (e.g. Cameron 1968). Three of the methods which have been proposed for oblique rotation are oblimax (Pinzka and Saunders 1954), oblimin (Carroll 1958), and promax (Hendrickson and White 1964). The oblimax program tends to run its axes a little bit outside the line of points that should be involved, while oblimin does the opposite. Promax seems to produce results similar to those of oblimax (Cattell and Khanna 1977).

Promax is one of the most widely-used and straightforward oblique transformations. It is based on an attempt to find the best least-squares fit between the oblique factor-pattern or factor-structure matrix and a target matrix which is thought to represent a simple structure solution. The target matrix is based on an orthogonal approximation to simple structure, usually a varimax factor loadings matrix. There are certain unusual situations in which the normal varimax procedure fails to arrive at an orthogonal approximation to oblique simple structure. In such cases the difficulties can be eliminated, or at least minimized, by using a weighted varimax rotation (see Cureton and d'Agostino 1983).

The theory of varimax and promax rotation was discussed in detail by Jöreskog *et al.* (1976) and Mather (1976). The former examined the geometry of oblique rotation, the latter gave FORTRAN computer programs for varimax and promax rotation, and for rotation to a specified target matrix.

Cattell (1965) pointed out that two objections can be raised to most rotation procedures: (i) They aim to prevent an attribute from having large loadings on two factors simultaneously, despite the fact that factors with similar loading patterns tend to occur in nature; (ii) The percentage of small loadings (below, say, 0.25) is only moderately correlated with the number of attributes actually in the hyperplane, yet the former function is made a substitute for the latter. Cattell preferred to consider a hyperplane passing through a q-dimensional cloud of points representing the attributes. Any points located in this hyperplane would have zero loadings on a 'reference vector' normal to it. The factor derived from this reference vector is then said to have the best possible simple structure position. Cattell and Khanna (1977) discussed simple structure rotation, and some problems which can be caused by clusters of attributes.

Maximization of the total number of attributes in the actual hyperplane is basic to Thurstone's original definition (Thurstone 1947) of the concept of simple structure. To fill this need, Cattell and Muerle (1960) developed the maxplane criterion, for which the program shifts hyperplanes and counts the number of points which are close to those hyperplanes in an attempt to find a configuration with the maximum number of attributes in the hyperplanes. Full details of the maxplane procedure are given in Cattell and Khanna (1977). Some analytical methods can rotate to positions which are often far from those which would be obtained by fitting hyperplanes, and so the maxplane criterion offers advantages. However, Katz and Rohlf (1974) found difficulty in convergence when using this technique on a number of data sets. Because the maxplane criterion is discrete, it is not possible to calculate smooth gradients. Also, 'micro-local maxima' may occur. In an attempt to overcome these problems, they developed the functionplane criterion, a continuous function which approximates the idea of maximizing a hyperplane count. The approach is to replace the simple hyperplane count with a non-linear weighted average of the distance of each attribute from a trial hyperplane. An exponential function is used so as to give greatest weight to attributes close to hyperplanes. Katz and Rohlf tested the method on a number of data sets and obtained solutions with clear hyperplanes. Computer times were comparable with those for varimax rotation.

Calculation of factor scores

Factor scores cannot be computed directly in the way that component values are calculated in a principal component analysis, although scores on image factors can be found exactly. The general factor analysis model (equation 4.1) is:

$$\mathbf{X} = \mathbf{FH}' + \mathbf{E}$$

In principal component analysis, $\mathbf{E}$ is assumed to be small, so the direct calculation of component values presents no problems. However, in factor analysis $\mathbf{E}$ is not assumed to be small. $\mathbf{F}$ and $\mathbf{H}$ are matrices concerned with the common part of $\mathbf{X}$, and $\mathbf{E}$ involves the unique parts of the attributes in $\mathbf{X}$. If there are q common factors, $\mathbf{F}$ will have q columns, whereas $\mathbf{E}$ will have p columns, one for each original attribute. We therefore have to estimate $p + q$ factor scores, not all of which are identifiable from p attributes alone. The matrix of loadings on the $p + q$ factors is not square and so does not have a unique inverse. Another problem is that factor scores cannot be estimated, in the statistical sense, because they are not population parameters.

Several formulae have been proposed for computing estimates of factor scores, perhaps the most commonly-used being Thurstone's regression-type least-squares solution. In Bartlett's method, minimizing the sum of squares of the standardized residuals (or scores on the specific factors) produces estimates which, unlike those of the least-squares method, are unbiased, but they have greater variability. Harris (1967) compared several methods of estimating factor scores, and he pointed out that the regression method yields scores which are intercorrelated even if the true factors are orthogonal. The strength of the correlation is a function of the number of attributes with a non-negligible loading on that factor. Such problems were also examined by McDonald and Burr (1967) and Tucker (1971).

Guttman (1955) found that for a given correlation matrix, factor loadings matrix, and data matrix, different factor score matrices can be postulated, i.e. there is a degree of indeterminancy. Yet another problem concerns computational accuracy. If the correlation matrix is nearly singular, inversion of the matrix, necessary in the computation of regression estimates of factor scores, can involve considerable error. Add to this the fact that the factor structure matrix is derived from the correlation matrix by an eigenvector transformation which itself may contain inaccuracies, and the possibility that computational errors can accumulate is evident.

Mather (1976) discussed in detail the calculation of factor scores, and gave FORTRAN computer programs for the regression method, Bartlett's method, and a method for estimation by 'ideal attributes', which is based on an approximation to the common parts of the attributes.

Interpretation

In factor analysis, interest centres mainly on the common factors, which are interpreted with reference to the observed or measured attributes. It is necessary to understand the difference between the factor structure matrix $\mathbf{H\Phi}$, which is used in the calculation of factor scores, and the factor pattern matrix $\mathbf{H}$ which contains the estimates of the parameters in the factor model. The factor pattern loadings are regression coefficients of the attributes on the factors, and are useful in interpreting the factors. They also give the contribution of each factor to the variances of the attributes. The loadings in the factor structure matrix are covariances between the attributes and factors. If the attributes are standardized, then these loadings are correlations. Both these matrices are required if the factors are oblique. Mather (1976) gave a geometrical explanation of the difference between the two matrices, see also Pimentel (1979 p98).

If the factors are orthogonal and standardized to unit variance, then $\mathbf{\Phi} = \mathbf{I}$ and both the factor pattern and factor structure are given by $\mathbf{H}$. The proportion of the variance of any attribute that is attributable to a common factor is the square of its loading on that factor, and the proportion of its variance that is attributable to its unique factor is one minus its communality. The sum of the squared elements of a row of $\mathbf{H}$ is the communality of the attribute represented by that row. The sum of the squared elements in a column is the variance along the corresponding factor. If the variates are standardized, dividing this value by p gives the proportion of the total variance accounted for by the factor. The communalities are invariant under orthogonal rotation, but the column sums of squares, and thus the proportion of the total variance accounted for by each factor, will change.

The loadings in the factor pattern matrix are the projections of the attributes onto the factors, and therefore a plot of the attributes in relation to the factor axes is a convenient way of displaying relationships among the attributes (e.g. see Jöreskog *et al.* 1976 p66). In the orthogonal case, the factor axes may be represented as two orthogonal vectors of unit length, in the oblique case the axes will not be at right angles.

The factor scores are important if the identification of patterns among the objects is an important aim. They are usually calculated with a mean of zero and unit variance, and can be used in ways similar to component values (see Chapter 5). Oblique factor scores are, of course, correlated.

Elffers (1980) discussed the interpretability of factor analysis results. He emphasized that the interpretation phase requires that we work within a theory powerful enough to enable us to decide whether a proposed interpretation is tenable or not. The theory may be rudimentary, but it must be made explicit so that it can be defended or rejected. He gave details of the calculation of the Guttman criterion for factor indeterminacy, although Mulaik (1976), Green (1976), and McDonald (1974, 1977) have argued against the use of this criterion. Elffers stated that if the Guttman criterion value of a factor makes it clear that all factor candidates, and thus all sets of factor scores, are very much alike, then it is not unreasonable to choose an arbitrary one or to use some approximation to all of them.

However, if the criterion suggests that the real differences between factor candidates are too large for justifiable interpretation, then we should not estimate factor scores at all. In either case, we need not concern ourselves about the difference between the methods for calculating factor scores.

When an initial exploratory factor analysis has been completed, it may be possible to develop a more specific model which can then be tested. This stage may be called confirmatory factor analysis. Darton (1980) mentioned two types of confirmatory factor analysis, the first involving maximum likelihood estimation of a factor model which includes restrictions specifying hypotheses concerning the parameters of the model, the second using target rotation. Darton also emphasized that, in general, too little attention has been paid to the problem of attribute selection in theory construction. These and other practical aspects of factor analysis were also discussed by Knight (1978).

There is no test for the significance of factor loadings. Sokal *et al.* (1961) arbitrarily selected a value of 0.4 as the lower limit for important loadings. Zelnio and Simmons (1981) agree with this value and added that an attribute should not correlate with any other factor at or above an absolute value of 0.3.

Harris (1975) discussed the interpretation of factor loadings and factor scores obtained using multiple regression. He pointed out that the contribution of a given attribute to the total predictability achieved by the set of attributes depends on what other attributes are included in the predictor battery. He suggested that in the process of naming (or reifying) factors, the major interpretative burden should be put on the coefficients used to derive the factor scores, rather than on the loadings. Indeed, the emphasis on factor score coefficients was one of Guttman's main motives for developing image analysis.

Stephens (1976) described a graphical method for displaying the positions of factor loadings or factor scores in relation to three factor axes. The projection simulates perspective by using circles to represent points and by decreasing the radius of the circle with distance from a selected viewpoint anywhere outside the space occupied by the axes.

In principal component analysis, the component values of objects can be used to study inter-object distances. In factor analysis the spatial relationships are slightly more complicated. Gower (1966, 1967b) noted that high correlations between sets of variates imply clusters of points, which is the basis of many of the rotation methods as an axis passing through the centre of one cluster and missing the centre of another will determine a factor which has nearly constant loadings for every variate within one cluster and zero loadings for variates in the other. Various methods have been proposed which combine cluster analysis with factor analysis (e.g. Thurstone 1947; Horst 1965; Tryon and Bailey 1970). In factor analysis, a cluster can be defined by a subset of attributes which have similar structures over all common factors, i.e. whose vectors are similarly oriented in the common-factor space. Examples are given in Cureton and d'Agostino (1983).

Applications of factor analysis

Factor analysis has not been widely used in ecology and biology. Dean and Gorham (1976) used factor analysis with varimax rotation on 43 chemical and mineralogical attributes of sediments from 46 Minnesota lakes. Four rotated factors accounted for 58% of the total variance and incorporated most of the relationships among the attributes. Hopke (1976) used maximum-likelihood factor analysis with varimax rotation on 32 chemical and physical attributes of lake sediments. Because of the poor statistical fit of this solution he tried a principal factor analysis. This yielded 5 factors with a similar pattern of coefficients to those obtained by the maximum-likelihood method.

Sokal *et al.* (1961) described a study which examined the value of factor analysis in a dynamic, rather than a static, biological model involving 19 biological and 6 physical attributes of insects and their environment. Part of the aim of this study was to include cause and effect attributes and to attempt to arrange them in order of causation. The authors concluded that the inclusion of possible causal attributes is to be highly recommended in exploratory studies. They also discussed the biological relevance of the concept of simple structure, and noted that although some people have misgivings about this concept, results of various researches have demonstrated that simple structure has invariably given a more meaningful interpretation to the data than has either a straight centroid solution or even a principal component analysis.

Delany (1965) used factor analysis on 10 attributes of 156 specimens of the long-tailed field mouse (*Apodemus sylvaticus* L.) from north-west Scotland. The analysis suggested the presence of three factors, and an oblique rotation was found to give a better fit than the orthogonal solution. The results suggested that life history and genetical size accounted for much of the variation, but there were also indications that animals from certain groups of localities were similar. The findings broadly agreed with those from a canonical variate analysis.

Nabholz and Richardson (1975) discussed the application of factor analysis to ecological problems, with special reference to mineral cycling. They emphasized the importance of the choice of attributes and the sample size. They commented that although common factors are useful in understanding and describing relationships within a scientific field, interpretation of the factors must be confirmed by evidence external to the factor analysis.

In recent years, there appears to have been an increasing use of factor analysis in microbiology. Rosswall and Kvillner (1978) drew the attention of microbiologists to this method and its possible applications. Unfortunately, they did not discuss the nature of the correlation matrix and the problems which arise with binary or multistate attributes. Väätänen (1980) used principal factor analysis with varimax rotation to study the influence of different environmental factors on aquatic microbial communities. About 90% of the communality of 16 microbiological attributes (with counts logarithmically transformed) was accounted for by 5 factors. Ten environmental attributes were also included. Väätänen concluded that the factor analysis had proved useful because it revealed that vernal bloom periods were important for the micro-organisms, a fact not shown by regression analysis. Bell *et al.* (1982) used principal factor analysis on physiological properties measured on continuous-culture enrichments of bacteria in two contrasting rivers. Eight separate analyses were performed, corresponding to combinations of summer–winter, river, plate-chemostat. Five major factors, emphasizing nitrogen metabolism, emerged regularly.

Holder-Franklin and Wuest (1983) found principal factor analysis with varimax rotation to be valuable in studies of population shifts in free-floating river bacteria. Two data matrices were used, one contained the results of 226 nutritional and physiological tests on isolates, the other contained 19 environmental attributes. The two data matrices were analysed separately and the factor scores from each analysis were cross-correlated. The analysis showed that water oxygen levels were significantly correlated with respiratory metabolism, nitrification, and lipid catabolism. Salt tolerance of the bacteria was positively correlated with the specific conductance and ion concentration of the water. Nitrate reduction was correlated with high carbon dioxide and low oxygen levels.

Confirmatory factor analysis can be used to test specific hypotheses about the causal structure between the underlying factors and the observed attributes. It could be a useful way of testing simple causal ecological models. Hypotheses to be tested may specify the number of common factors involved, whether they are orthogonal or oblique, and the magnitudes of the factor loadings. Such hypotheses could be based on preliminary ideas about the causal mechanisms. A program for confirmatory factor analysis was given in Sörbom and Jöreskog (1976, see also Kim and Mueller 1978 p 46 *et seq.*). This enables two types of analysis to be carried out: (i) the test of a hypothesis for one group of objects, and (ii) the comparison of hypothetical factor structures across several groups of objects.

Advantages and disadvantages of factor analysis

Proponents of factor analysis (e.g. Cattell 1965) point out that it is unlikely that a small sample of attributes will actually represent the real common influences required to account for all the variance of an attribute. Indeed, we may not know which attributes we should measure, even if we wanted to measure all of them and could. From this point of view, it does seem reasonable to explore the possible uses of factor analysis in environmental studies. A second argument is that, in nature, the underlying causal influences which we are seeking are unlikely to be completely independent. The factor analysis model specifically allows for the possibility of correlated factors. If, as in principal component analysis, orthogonal factors are calculated where the basic causal factors are really correlated, the elements of the factor loadings matrix will be difficult to interpret.

The principal component solution is not a particularly parsimonious one, because to account for all the variance, r components are needed, where r is the rank of the correlation or covariance matrix. Dimensionality can only be reduced further by discarding some components. This is done at the expense of possible systematic error in the reproduction of the original intercorrelations, as there may be one or more attributes which are more highly related to the discarded components than to those retained. The effect of replacing the values of unity in the diagonal of a correlation matrix by values less than unity (the communality estimates) is to reduce the determinant of the correlation matrix. Factor analysts argue that the apparently higher rank of the initial correlation matrix was an artefact caused by including in the analysis error variance and specific variance, neither of which can be involved in the inter-relationships among the attributes. What is lost by using factor analysis as opposed to principal component analysis is the straightforward relationship between the values of the attributes for an object and its component values. Estimation of the scores of the objects on a given factor axis requires a multiple regression analysis of the relationship between the factor and the attributes, with the multiple correlation coefficient of the estimates being a decreasing function of the amount of unique variance in the system (Harris 1975). Guttman's image analysis overcomes the indeterminacy of the relation between the factors and the attributes.

A second loss in most factor analysis methods is the uniqueness of the solution. The original pattern of intercorrelations is represented by a given factor structure, but that pattern could be described equally well by a range of other orthogonal solutions using the same number of dimensions. Principal factor analysis assures the uniqueness of the factor structure that it produces by making each successive factor account for the maximum possible proportion of the common variance. The maximum-likelihood method generates factors so as to maximize the probability that the observed pattern of correlations could have arisen through random sampling from a population in which the correlations are perfectly reproducible from the number of factors specified by the user (Harris 1975). The Minres method produces factor loadings which give the smallest possible sum of squared residuals between the observed and reproduced correlations.

Another important disadvantage of factor analysis is that not only is it computationally more complicated than is principal component analysis, but it also places a much heavier demand on the user, who must have some understanding of the theory and some knowledge of the system under study, otherwise meaningless results could be accepted uncritically. Mather (1976) discussed three practical aspects of factor analysis:

(i) The problem of deciding the number of factors required. In principle, enough factors should be extracted so that the matrix of residual correlations is not significantly different from an identity matrix, although there is no satisfactory way of testing this except in maximum-likelihood factor analysis, which requires an assumption of normality. The chi-square test has been criticized because it depends on the number of objects, and it may also lead to over-factoring.

(ii) Normality of observed variates is not an explicit requirement except in maximum-likelihood methods. However, the interpretation of the factor pattern may be affected by non-normality, since the sizes of the factor pattern coefficients are influenced by the sizes of the correlation coefficients, which may be restricted if the distributions of the original variates are highly skewed (Carroll 1961).

(iii) There is difficulty in deciding whether or not a factor pattern coefficient is significant (a similar problem occurs in principal component analysis). Mather pointed out that in a 50×10 factor pattern matrix, we could expect to accept at least 25 entries as significantly different from zero at the 5% level even if all the population factor pattern coefficients were really zero. Replication may help to avoid errors of this type.

Factor analysis has the same assumption as does principal component analysis that any relationships among the original variates are essentially linear. Furthermore, as in principal component analysis, the derived factors are linear combinations of the original variates. McDonald (1962, 1967a,b) discussed non-linear factor analysis. Marriott (1974) suggested that the linearity constraint is likely to provide a reasonable approximation if the range of attributes is very small, but is not likely to work well for extreme values.

Pimentel (1979) was not happy about the concept of simple structure, and gave two biological reasons for rejecting it. Firstly, biologists are interested in the various ramifications of a phenomenon and patterns of variation, rather than in static conditions. Secondly, biological data rarely sufficiently circumscribe a phenomenon to allow the simple structure discovered to be that of the phenomenon. Nevertheless, various researchers, discussed above, have found it useful.

So, should a researcher try the application of factor analysis, and if so, in what circumstances? Factor analysis should be regarded as an alternative to principal component analysis. With some data one method will give a more interpretable result, whereas with other data the alternative method will prove better. Principal component analysis will always yield a result, but it will not necessarily be possible to interpret the components. Proponents of factor analysis claim that the separation of the total variance into common and unique variances gives factors which are likely to be easier to interpret, and are probably more realistic. If the communalities of the attributes are all equal, then there is probably nothing to be gained by using factor analysis rather than principal component analysis (e.g. see Harris 1975). Harris commented that in using factor analysis, the attributes should be nearly homogeneous with respect to their reliabilities (specific variances) but heterogeneous with respect to their communalities. Then, if accurate communality estimates are obtained, relatively more of the total error variance in the system will be retained in the first q factors than in the first q components.

Clifford and Stephenson (1975) were optimistic about the usefulness of factor analysis in ecology. They cited as one example the case in which the data include a large proportion of attributes with only weak intercorrelations, e.g. as might occur with certain species of low frequencies. With standardized data, each species would contribute equally to the total variance, and the proportion of the total variance extracted by the larger components would be small (in fact, it would be inversely proportional to the number of such species). In factor analysis, on the other hand, weakly-correlated species-pairs play only a minor part in determining the magnitudes of the communalities.

Gower (1967b) commented that whether or not factor analysis is likely to be helpful depends on the strength of the belief in the existence and interpretation of the specific factors.

Finally, Lawley and Maxwell (1971 p38) commented that it should always be kept firmly in mind that except in artificial experiments the basic factor model, like other models, is useful only as an approximation to reality, and it should not be taken too seriously.

CHAPTER 7

PRINCIPAL CO-ORDINATE ANALYSIS

Background

Principal co-ordinate analysis is a computationally-simple yet powerful procedure which has wide applications. It arose from a dissatisfaction with many reported applications of factor analysis and principal component analysis of Q-mode matrices in classification studies, particularly in the biological literature. If many, or all, of the attributes are qualitative, product-moment correlations between attributes may be inappropriate. In such a case, the analysis requires the use of a Q-mode matrix of similarity coefficients, or of some function of a similarity coefficient which is regarded as the distance between two objects. Although the analysis of such a matrix is a Q-mode analysis, some analysts applied the techniques of principal component analysis and factor analysis to such matrices and, in spite of the fact that in such an analysis the standard underlying assumptions are not even approximately satisfied, obtained results which were successful in that the expected relative magnitudes of interobject distances were recovered.

Suppose we have a data matrix $\mathbf{X}$ ($N \times p$). Any row vector $\mathbf{x}'_i$ gives the co-ordinates of a point P_i which represents the position of an object in a p-dimensional space. From $\mathbf{X}$ we can derive a matrix, of order N, of coefficients of similarity between objects. The eigenvectors of this matrix give the co-ordinates of the points P_i ($i = 1$ to N) on the new axes. The problem is, what is the correct scaling for each eigenvector, and what is the interobject distance when this scaling is used? This problem was solved by Gower (1966), and a rather simplified account is given in Gower (1967b).

Theory

Suppose we have a symmetric matrix $\mathbf{A}$, of order N, the elements of which are some form of similarity coefficient. Matrix $\mathbf{A}$ has N eigenvalues and associated eigenvectors $\mathbf{c}_j$, which are the columns of $\mathbf{C}$. The elements of the ith row of $\mathbf{C}$ are taken as the co-ordinates of a point P_i in Euclidean

107

N-space. The Pythagorean distance between two points P_i and P_j in this space is

$$d_{ij} = \left[\sum_{k=1}^{N} (c_{ik} - c_{jk})^2 \right]^{1/2} \tag{7.1}$$

Gower (1966) showed that if the eigenvectors are normalized so that the sum of squares of an eigenvector is equal to the corresponding eigenvalue, then the distance $\Delta (P_i, P_j)$ is

$$[\Delta(P_i, P_j)]^2 = a_{ii} + a_{jj} - 2a_{ij} \tag{7.2}$$

If the similarity coefficient is scaled from zero to a maximum of unity for complete identity, then the diagonal elements of **A** will be unity, and the interpoint distance in equation (7.2) becomes

$$[\Delta(P_i, P_j)]^2 = 2(1 - a_{ij}) \tag{7.3}$$

Clearly, the distance $\Delta(P_i, P_j)$ will be zero for complete identity and will reach a maximum for total dissimilarity.

The validity of these relationships depends upon matrix **A** fulfilling two conditions. First, it must be symmetric, otherwise the relationship in equation (7.1) does not hold. Fortunately, most of the commonly-used similarity measures (Chapter 12) are symmetric, although Williams, Dale and Lance (1971) gave an example which failed in this respect. The second requirement is for the matrix from which the eigenvalues and eigenvectors are calculated to be positive semi-definite, otherwise one or more of the eigenvalues will be negative and part of the space will be imaginary, and therefore non-Euclidean. As N points can be fitted into $N - 1$ dimensions, there will always be at least one zero eigenvalue. Gower (1966) showed that the transformation

$$b_{ij} = a_{ij} - m_i - m_j + g \tag{7.4}$$

gives a matrix **B** which is positive semi-definite for most of the commonly used similarity measures, although not for an information-fall statistic (Williams *et al.* 1969). Here, m_i and m_j are the means of the ith and jth rows (or columns) of **A** and g is the grand mean. This transformation does not alter the interpoint distances.

Duality with principal component analysis

The proof of the duality with principal component analysis is due to Gower (1966). If a data matrix $X(N \times p)$ contains continuous variates which are expressed as deviations from the variate means, the minor product moment matrix $X'X$ is of corrected sums of squares and cross-products, and can be used in a principal component analysis. Let the eigenvalues of this matrix be λ_j and the eigenvectors be $\mathbf{u}_j$ and

$$\mathbf{X'Xu}_j = \lambda_j\mathbf{u}_j \tag{7.5}$$

We know that the eigenvalues of a major product moment matrix are the same as those of the corresponding minor product moment matrix, so that

$$\mathbf{XX'v}_j = \lambda_j\mathbf{v}_j \tag{7.6}$$

Premultiplying (7.5) by $\mathbf{X}$ we get

$$\mathbf{XX'}(\mathbf{Xu}_j) = \lambda_j(\mathbf{Xu}_j) \tag{7.7}$$

Hence the two sets of eigenvectors are related by a scaling constant k, i.e. $\mathbf{Xu}_j = k\mathbf{v}_j$. In principal component analysis, the eigenvectors are normalized so that $\mathbf{u}_j'\mathbf{u}_j = 1$. From the above principal co-ordinates analysis theory $\mathbf{v}_j'\mathbf{v}_j = \lambda_j$, so that

$$k^2\mathbf{v}_j'\mathbf{v}_j = \mathbf{u}_j'\mathbf{X'Xu}_j = \lambda_j\mathbf{u}_j'\mathbf{u}_j = \lambda_j \tag{7.8}$$

Hence, for this scaling k = 1 and

$$\mathbf{v}_j = \mathbf{Xu}_j \tag{7.9}$$

which means that the co-ordinate values are the same as the component values.

If $\mathbf{A} = \mathbf{XX'}$ then

$$a_{ii} = \sum_{k=1}^{p} x_{ik}^2 \quad \text{and} \quad a_{ij} = \sum_{k=1}^{p} x_{ik}x_{jk}$$

so that equation (7.2) becomes

$$[\Delta(\mathrm{P}_i, \mathrm{P}_j)]^2 = \sum_{k=1}^{p} (x_{ik} - x_{jk})^2$$

which is the Pythagorean distance between the two points on the original variates. The same solution can be obtained if the elements $a_{ij} = -0.5d_{ij}^2$, where d_{ij} is as given by equation (7.10). Then, substituting in equation (7.2)

$$[\Delta(\mathrm{P}_i, \mathrm{P}_j)]^2 = d_{ij}^2$$

because $d_{ii} = 0$ for $i = 1$ to N.

There are two special cases of this duality. The first concerns standardized variates $z_{ij} = x_{ij}/s_j$, where the x values are expressed as deviations from the sample means. Then $\mathbf{Z'Z} \times 1/(N-1)$ is a correlation matrix, with eigenvalues λ_j and components with sums of squares equal to $\lambda_j(N-1)$. The eigenvalues of $\mathbf{ZZ'}$ (and $\mathbf{Z'Z}$) are $N-1$ times those of the corresponding correlation matrix, i.e. are equal to $\lambda_j(N-1)$. The co-ordinates obtained from $\mathbf{ZZ'}$ are made to have sums of squares equal to the corresponding eigenvalues, hence they are the same as the components obtained from the corresponding correlation matrix. Furthermore, equation (7.10) becomes:

$$[\Delta(P_i, P_j)]^2 = \sum_{k=1}^{p} (z_{ik} - z_{jk})^2$$

which is the Pythagorean distance on the standardized variates, and is also Sokal's taxonomic distance measure (Sneath and Sokal 1973 p124).

The second special case concerns binary (0, 1) data. In the usual notation for similarity coefficients (Chapter 12) the simple matching coefficient is

$$S_{ij} = \frac{a + d}{p}$$

Clearly, S_{ij} = zero if there are no matches, and $S_{ii} = 1$ for complete identity. Using such a matrix for **A**, equation (7.10) becomes

$$\sum_{k=1}^{p} ((x_{ik} - x_{jk})^2 = b + c = p(1 - S_{ij})$$

Therefore, a principal component analysis of the matrix of corrected sums of squares and products between binary attributes is equivalent to assuming that the objects are represented by points whose distances apart are proportional to $\sqrt{(1 - S_{ij})}$.

Choice of attributes and type of data

The comments on the choice of attributes are essentially the same as those given for principal component analysis. Although it may be convenient to use whatever data are available, or are easily measured, wherever possible the attributes should be chosen because they are considered to be relevant, or because the researcher wishes to examine relationships for a particular purpose.

The user of a principal co-ordinate analysis has a greater choice of data types than is possible with principal component analysis. Gower's (1971a) general similarity coefficient can accommodate mixtures of binary, multi-state, and quantitative attributes. For purely binary data a range of similarity coefficients is available (see Chapter 12).

It should be noted that the only theoretical requirement for principal co-ordinate analysis is that matrix **B** should be symmetric and positive semi-definite, and this will be true for most of the commonly-used similarity measures. However, missing values in the original data matrix may destroy this property (e.g. see Gower 1971a). The space spanned by the principal co-ordinate axes is Euclidean even if the original attribute space is not.

Computational procedure

(i) Form the association or similarity matrix **A**. Perhaps the most important aspect of this type of analysis is the choice of a suitable measure, this is discussed in Chapter 12.

(ii) Transform **A** to **B** by equation (7.4).

(iii) Calculate the eigenvalues d_j and eigenvectors $\mathbf{v}_j$ of matrix **B**, the eigenvectors being normalized so that the sum of squares of each eigenvector is equal to the associated eigenvalue d_j, i.e. $\mathbf{v}_j'\mathbf{v}_j = d_j$. The columns of **V** are the 'Gower vectors', which give the co-ordinates of the points (objects) on the new axes.

It is clear that it is very easy to run into practical computing problems in calculating the eigenvalues and eigenvectors of a matrix of order N when N is large, both time and storage space, especially space, can become a problem. One way of overcoming this problem is to use a procedure which first tridiagonalizes the matrix using an algorithm which needs only the lower half-matrix to be stored, e.g. the TRED3 algorithm of Martin *et al.* (1971). As only a few of the largest eigenvalues will be required, time can be saved by using an algorithm written specially for this purpose (e.g. the RATQR algorithm of Reinsch and Bauer 1971). The corresponding eigenvectors of the tridiagonal matrix can then be calculated by using the inverse iteration procedure in algorithm TRISTURM of Peters and Wilkinson (1971), and these eigenvectors are finally back-transformed to those of the original matrix using algorithm TRBAK3 (Martin *et al.* 1971).

However, even with these procedures memory space can become a problem. The number of elements in a similarity half-matrix with the principal diagonal omitted is $N(N - 1)/2$, so that for 50 objects there are 1225 elements and for 100 objects there are 4950 elements. If the number of objects is large enough to cause memory problems, one possibility is to use an algorithm which does not require the half-matrix to be stored (e.g. Paige 1972).

Another approach, suggested by Lefkovitch (1976), uses the relationship between the eigenvalues of $\mathbf{X'X}$ and $\mathbf{XX'}$ which is given by the singular value decomposition (Section 2.3). Lefkovitch distinguished two cases by considering whether $\mathbf{X'X}$ has meaning (case 1) or not (case 2). In case 1, the columns of **X** are continuous variates which are linearly (or at least monotonically) related. If the attributes are not centred on the means, the addition of an attribute which is unity for all objects, before computing $\mathbf{X'X}$, will take care of the centring. Then compute the eigenvalues and eigenvectors of $\mathbf{X'X}$ and obtain the principal co-ordinate values using equation (7.9), where $\mathbf{u}_j$ are the eigenvectors in order of decreasing eigenvalues. Lefkovitch pointed out that it is possible to transform multistate attributes so that $\mathbf{X'X}$ has meaning. In case 2 the analysis is of the similarity or distance matrix as usual.

Yet another possibility was suggested by Gower (1968). If a principal co-ordinate analysis of a base set of N points is already available, it is quite straightforward to calculate the co-ordinates of a new $(N + 1)$th point relative to the axes of the base set. A practical example of this method in taxonomy was given by Wilkinson (1970). This method could be used for adding many new points, but the result would then be approximate (see Ross 1969a).

Dimensionality

If the kth eigenvalue is small, then the contribution of $(v_{ik} - v_{jk})^2$ to the total interpoint distance d_{ij} will be small. If the kth eigenvalue is large but the elements of the corresponding co-ordinate axis are not very different, then again the contribution to the interpoint distance will be small. Therefore, the only co-ordinates which contribute much to the interpoint distances are those which have large eigenvalues and also show a large variation in the sizes of the co-ordinate values. In many applications it is found that the interpoint distances can be adequately expressed in terms of two or three such co-ordinate vectors.

Reification

Principal co-ordinate analysis has more limited aims than do principal component analysis and factor analysis. In the former, the main aim is the graphical treatment of the data. As N points can be fitted into $N - 1$ dimensions, principal co-ordinate analysis must yield at least one zero eigenvalue. If matrix $\mathbf{B}$ is positive semi-definite, the r non-zero eigenvalues will be positive. The eigenvectors corresponding to the q largest positive eigenvalues give the co-ordinates of points Q_i which are the projections of the original points P_i on the best-fitting subspace of dimensionality q. The sum of squares of the values on a co-ordinate axis (eigenvector) is equal to the corresponding eigenvalue by definition. As in principal component analysis, the sum of squares of the residuals will be the difference between the trace of matrix $\mathbf{B}$ and the sum of its q largest eigenvalues.

The first co-ordinate axis maximizes the total squared distance among the objects, the second minimizes the total squared distance along an axis orthogonal to the first, and so on. The positions of the points can be plotted on pairs of rectangular co-ordinate axes, and the resulting ordination chart can be examined for pattern. Because

$$\mathbf{B} = \mathbf{VV'}$$

any diagonal element b_{ii} is the squared distance of point Q_i from the origin O. Furthermore, an off-diagonal element b_{ij} is the cosine of the angle between the vectors from the centroid O to Q_i and Q_j times the product of distances OQ_i and OQ_j. The better the approximation given by the first q vectors, the closer will Q_i be to P_i (Jöreskog *et al.* 1976).

The interpretation of the axes is more complicated than in principal component analysis. It involves calculating the relationships between the values of the points on a co-ordinate axis and on each of the original attributes. Each such relationship can be expressed as a pseudo F statistic, and although these values cannot be treated statistically they serve to rank the attributes in order of the importance of their contributions to a co-ordinate axis.

If matrix **A** consists of measures which do not have a Euclidean representation, matrix **B** may not be positive semi-definite, and therefore may have negative eigenvalues. Provided that these are sufficiently small they may, in practice, be ignored. Cailliez and Pages (1976) showed that the Euclidean reduced-space approximation obtained by principal co-ordinate analysis is acceptable if the largest negative eigenvalue is not larger in absolute value than any of the positive eigenvalues of interest. With negative eigenvalues, the quality of the reduced-space representation is given by

$$\frac{\sum_{i=1}^{q} \lambda_i + q|\lambda_p|}{\sum_{i=1}^{p-1} \lambda_i + (p-1)|\lambda_p|}$$

where q is the number of positive eigenvalues of interest, p is the total number of dimensions, and $|\lambda_p|$ is the absolute value of the largest negative eigenvalue.

Uses of principal co-ordinate analysis

Principal co-ordinate analysis is useful as a general ordination technique if the data are multistate attributes, or are mixtures of binary, multistate and quantitative attributes. In this application, it is a useful technique for producing plots on which to examine the results of a cluster analysis applied to a similarity or distance matrix computed directly from the data.

Principal co-ordinate analysis has been used in the taxonomy of organisms (e.g. Corbet *et al.* 1970; Wilkinson 1970; Blackith and Reyment 1971), and in geology (e.g. Reyment *et al.* 1976). Rayner (1966, 1969) used principal co-ordinate analysis to compare soils in Glamorganshire (Wales). This approach gave a very clear interpretation along two axes, which were interpreted tentatively as reflecting acidity and redox potential. The major soil types involved were clearly separated on a two-dimensional scatterplot. Although the result was similar to that of a principal component analysis, the principal co-ordinate solution was considered to be easier to interpret.

Its main application in ecology seems so far to have been with binary data as a dual to principal component analysis, and like principal component analysis it can be used as a preliminary orthogonalizing transformation. Williams (1976) gave some interesting examples in which the co-ordinate values have been used for conventional statistical analysis. Although such use has aroused misgivings among statisticians, this application does appear to be effective.

Howard and Howard (1988) used Gower's general similarity coefficient in the principal co-ordinate analysis of mixed quantitative, multistate, and binary attributes obtained from maps. The resulting Euclidean co-ordinates were then subjected to *k*-means clustering.

Williams (1976) also described a technique which he called canonical co-ordinate analysis. In the normal canonical correlation of two batteries of ecological attributes measured on the same objects, the canonical axes obtained are often difficult to interpret, possibly because there is a high proportion of unique variances which confuses the common factor pattern. In canonical co-ordinate analysis, the two batteries of attributes are separately subjected to principal co-ordinate analysis, and from each set of co-ordinates those associated with the largest eigenvalues are selected and used in a canonical correlation analysis. Because of the preceding principal co-ordinate analysis the computations are simplified, and examples have shown that interpretation is also simplified. Examples are given in Williams (1976).

Lefkovitch (1976) described a computationally-simple divisive method for hierarchical clustering. He showed that the co-ordinate vectors, in decreasing order of their eigenvalues, indicate the successive levels of the hierarchy of a dendrogram, and the signs of the vector elements indicate the group membership. There are three assumptions in this method: (i) that matrix **A** (or **XX'**) is an appropriate description of the interobject relationships; (ii) that the objects do, in fact, belong to disjoint, rather than overlapping, groups; (iii) that a group formed by co-ordinate vectors 1 to j is either unchanged by including co-ordinate vector $j + 1$, or it is divided into two. If these assumptions hold then this method maximizes the between-group squared distance.

Advantages and disadvantages of principal co-ordinate analysis

Principal co-ordinate analysis is especially useful when only similarities or distances are known, for example as in certain biochemical tests. Mather (1976) noted that principal co-ordinate analysis is computationally more efficient than is multidimensional scaling or non-linear mapping, because an analytical rather than an iterative procedure is employed. It is perhaps worth noting here that methods which begin with interpoint distances and then derive co-ordinates on which the points can be plotted, are referred to as scaling methods. Principal co-ordinate analysis of a symmetric matrix of $-0.5\,\Delta^2$ is the same as the classical scaling of Young and Householder (Greenacre and Underhill 1982). These are related to other scaling methods such as correspondence analysis.

An important advantage of principal co-ordinate analysis is that it can be used for quantitative, binary, or multistate attributes or mixtures of them.

In principal component analysis, the interobject distances on the observed variates are assumed to be Pythagorean. An advantage of principal co-ordinate analysis is that the original interobject distances do not need to be Euclidean, but even if they are not the co-ordinate space (or 'Gower-space') is Euclidean. The user is free to choose a similarity measure (subject to the matrix being symmetric and positive semi-definite), and in so doing can insert into the analysis some ideas about aspects of the data which are thought to be important. For example, one may wish to emphasize species dominance and hence use the Bray–Curtis measure. Alternatively, one may be more interested in relative properties and so use the Canberra metric (Clifford and Stephenson 1975).

Another point to bear in mind is that in practice a principal component analysis is usually performed on standardized data, so that all attributes are equally important in their contribution to the total variance. In principal co-ordinate analysis the user has more choice in this respect.

Another advantage is that some types of similarity measure are reasonably robust and reliable if there are missing values, whereas replacing the missing values by estimates or guesses is not very satisfactory (Marriott 1974).

The main disadvantage of principal co-ordinate analysis seems to be non-linearity with some plant species presence–absence data (e.g. Ingwersen 1983), a problem which it shares with principal component analysis.

A practical application of principal co-ordinate analysis

Table 7.1 gives values for seven quantitative attributes, one multistate attribute, and one binary attribute for 30 male and female swans, half of which were sampled from a Polish population and half from a British one (Dr. P.J. Bacon, personal communication). Gower's general similarity coefficients between swans were computed, and the eigenvalues of the adjusted similarity matrix are given in Table 7.2. As the matrix is of order 30 there is a maximum of 29 non-zero eigenvalues and corresponding eigenvectors. In principal co-ordinate analysis it is usually necessary to

Table 7.1 Data for 30 male and female swans from British and Polish populations

Swan no.	Weight kg	Skull length mm	Radius length mm	Tarsus length mm	Web width mm	Bill knob width mm	Length of 3rd primary feather mm	Moult score	Sex M = 0 F = 1
1	9.1	169	285	109	179	12	0	0	1
2	13.0	179	328	117	197	20	365	9	0
3	12.4	184	303	111	187	23	356	9	0
4	9.5	167	279	102	166	15	48	1	1
5	12.6	179	295	116	178	15	0	0	0
6	11.0	185	314	112	188	13	362	5	0
7	13.2	179	322	119	182	14	374	9	0
8	7.6	163	388	107	156	11	82	2	1
9	8.1	168	282	101	159	11	149	2	1
10	8.7	169	283	104	167	10	317	9	1
11	8.6	172	279	104	165	11	328	9	1
12	10.1	182	312	113	186	12	348	9	0
13	10.8	184	303	112	182	16	360	9	0
14	9.6	169	282	107	170	14	142	2	1
15	11.9	187	318	119	177	13	64	1	0
16	7.5	170	282	114	159	12	177	3	1
17	8.9	180	284	111	158	10	350	9	1
18	11.3	193	282	114	175	15	268	4	1
19	11.5	184	298	115	165	16	193	2	0
20	12.0	187	305	121	174	20	130	2	0
21	7.4	173	270	106	165	15	165	2	1
22	9.8	181	276	112	177	16	342	9	1
23	8.0	164	255	104	168	13	98	1	1
24	10.8	186	290	107	180	19	100	2	0
25	8.4	182	273	109	171	13	129	1	0
26	11.0	185	290	113	190	14	219	3	0
27	7.8	161	260	99	152	13	40	1	0
28	9.0	176	270	106	160	17	139	2	0
29	10.0	181	300	112	174	15	50	2	0
30	9.6	165	271	119	155	18	290	4	1

Numbers 1 to 15 are from Poland, 16 to 30 are from Britain.

compute only a few of the largest eigenvalues, but in this example all the eigenvalues have been listed to illustrate their distribution.

As was explained earlier, there is no simple rule-of-thumb for interpreting the eigenvalues, but in general the only axes which contribute much to the interobject distances are those which have large eigenvalues and also have a large variation in the sizes of the co-ordinate values. In this example, we might reasonably conclude that most of the useful information is contained in the first two axes, which together account for 47% of the variation.

The pseudo F statistic, which serves to rank the attributes in order of the importance of their contributions to the co-ordinate axes, showed that the attributes which contributed most to the first axis were weight, sex, skull length, and web width in that order. The attributes which contributed most to the second axis were length of third primary feather and moult

Table 7.2 Eigenvalues of the adjusted similarity matrix for the swan data in Table 7.1

		Percentage of trace	
Axis	Eigenvalue	Axis	Cumulative
1	3.3979	30.61	30.61
2	1.8161	16.36	46.98
3	0.9868	8.89	55.87
4	0.7749	6.98	62.85
5	0.5375	4.84	67.69
6	0.5191	4.68	72.37
7	0.4660	4.20	76.57
8	0.3464	3.12	79.69
9	0.2916	2.63	82.32
10	0.2766	2.49	84.81
11	0.2024	1.82	86.63
12	0.1967	1.77	88.41
13	0.1753	1.58	89.98
14	0.1583	1.43	91.41
15	0.1314	1.18	92.59
16	0.1032	0.93	93.52
17	0.1012	0.91	94.44
18	0.0895	0.81	95.24
19	0.0810	0.73	95.97
20	0.0725	0.65	96.63
21	0.0666	0.60	97.23
22	0.0521	0.47	97.70
23	0.0495	0.45	98.14
24	0.0477	0.43	98.57
25	0.0447	0.40	98.97
26	0.0363	0.33	99.30
27	0.0314	0.28	99.58
28	0.0262	0.24	99.82
29	0.0200	0.18	100.00

score. A scatterplot of points representing the swans on the first two axes is given in Figure 7.1. All but three of the points with positive values on the first axis represent female swans, the exceptions being the smallest male swans numbers 25, 27 and 28. All but two of the points with negative values on the first axis represent male swans, the exceptions being the largest female swans numbers 18 and 22. This result is consistent with the known sexual dimorphism. The nine points with the largest positive values on the second axis represent swans with third primary feather lengths of 317 mm or more and moult scores of 9. The four points with the smallest positive values on the second axis, and one with a small negative value, represent swans with third primary feather lengths of 177 to 362 mm and moult scores of 3, 4 or 5. The remaining points with negative values on the second axis have third primary feather lengths of 0 to 193 mm and moult scores of 0, 1, or 2.

The first two axes did not distinguish between swans from British and Polish populations.

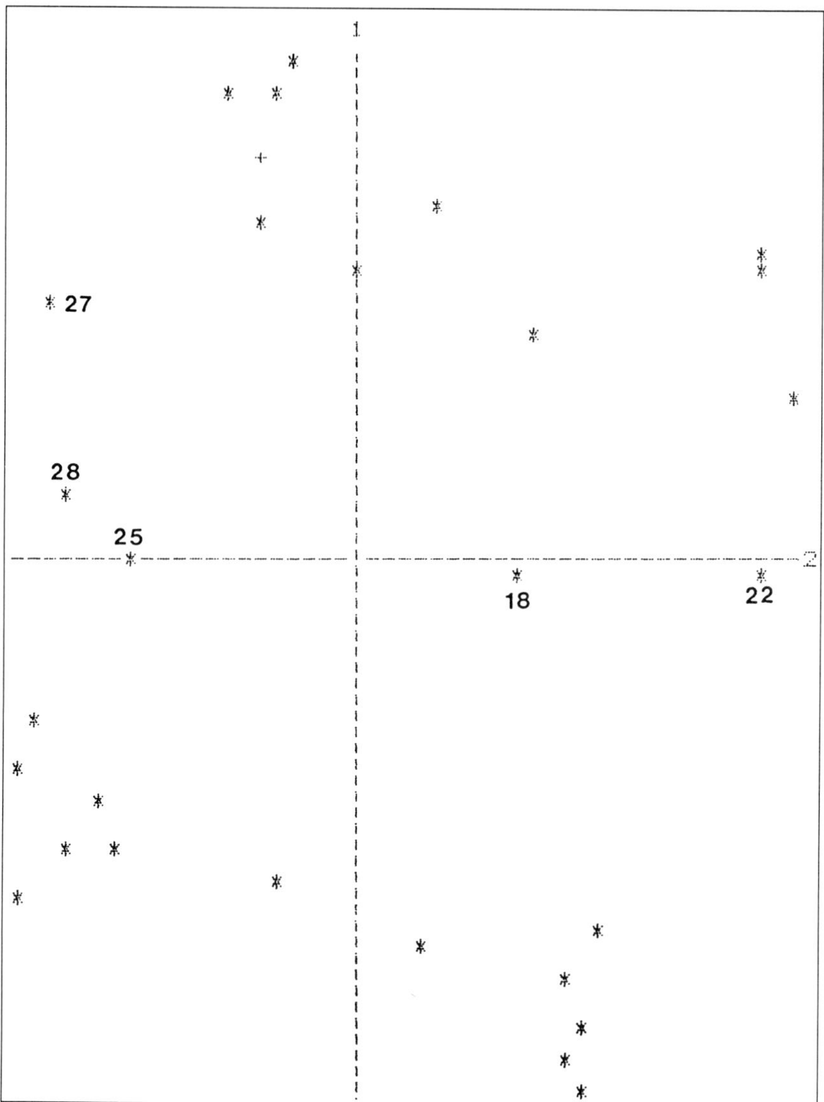

Figure 7.1 Scatterplot of points, representing the 30 swans, on the first two co-ordinate axes (+ represents coincident points 4 and 21)

If it were considered to be useful to perform a cluster analysis on such data, a hierarchical method could be used on the original similarity matrix or derived distances, or a non-hierarchical method such as k-means could be used on the co-ordinate values.

CHAPTER 8

Q-MODE FACTOR ANALYSIS

The main use of Q-mode factor analysis has been in geology (Imbrie and Purdy 1962; Imbrie 1963; Manson and Imbrie 1964; Imbrie and van Andel 1964). The method is based on the idea that the composition of each object (geological specimen) in a data matrix could result from the mixing, in various proportions, of several hypothetical objects of specific, but unknown, composition. The aims of the analysis are therefore: (i) to find the minimum number of hypothetical objects, or assemblages of them, of which the observed objects may be considered to be combinations; (ii) to specify the compositions of the hypothetical objects; (iii) to describe each observed object in terms of the hypothetical objects.

In effect, Q-mode factor analysis consists of finding the principal axes of a q-dimensional hyperellipsoid which is defined by the interrelationships between N object vectors. As each object is defined by p attributes, the true dimensionality of the problem can be no greater than p. The factors obtained may be considered to represent the hypothetical objects, which are idealized end-members of the collection of observed objects. Alternatively, the factors may be rotated until each one coincides with an observed object which then becomes an end-member. Davis (1973) noted that, if an oblique rotation is used, the factors, and hence the end-members, will be intercorrelated. This may not be a problem if the purpose of the analysis is simply to arrange the objects on a gradient, but it would be if we wished to draw inferences outside the data set.

Theory

The theory is given, with worked examples, in Jöreskog *et al.* (1976) and Davis (1973). The first step is to form a similarity matrix of order N. The definition of similarity is central to the analysis, but however defined it must lie within the range -1 to 1. Imbrie and Purdy (1962) defined the cosine θ coefficient.

$$\cos \theta_{ij} = \frac{\sum_{k=1}^{p} x_{ik} x_{jk}}{\sqrt{\sum_{k=1}^{p} x_{ik}^2 \sum_{k=1}^{p} x_{jk}^2}} \tag{8.1}$$

119

This is the cosine of the angle between the *p*-dimensional row vectors of the data matrix $\mathbf{X}$ ($N \times p$) which represents objects *i* and *j*. It is also an index of proportional similarity as well as being a measure of the great circle distance between points *i* and *j* which lie on the surface of a hypersphere. As it defines proportional similarity, this coefficient takes no account of size, so a large object and a small object may be considered to be identical if they have the same proportional composition. The cosine θ coefficient is equal to the product-moment correlation coefficient if the variates have zero means and unit variances. As the concept of 'variances' between variates is not appropriate, the product-moment correlation coefficient is not recommended for this purpose.

In matrix notation, the similarity matrix $\mathbf{C}$ ($N \times N$) is given by

$$\mathbf{C} = \mathbf{WW'} \tag{8.2}$$

where $\mathbf{W}$ is the row-normalized data matrix

$$\mathbf{W} = \mathbf{D}^{-1/2}\mathbf{X} \tag{8.3}$$

$\mathbf{D}$ ($N \times N$) is a diagonal matrix of the row sums of squares of $\mathbf{X}$, and the row vectors in $\mathbf{W}$ are of unit length. Row normalization removes the effects of size differences between objects without affecting the proportionality relations between their attributes. From the general factor analysis model (equation 4.1), and assuming that the error terms are small

$$\mathbf{W} = \mathbf{HF'} \tag{8.4}$$

where $\mathbf{H}$ ($N \times r$) is a factor loadings matrix, $\mathbf{F}$ ($p \times r$) is a factor scores matrix, and *r* is the rank of $\mathbf{W}$.

The first of the aims listed above, i.e. to find the minimum number of end-members, can be achieved by finding a lower-rank approximation to $\mathbf{W}$. If a *q*-dimensional approximation to $\mathbf{W}$ is sufficiently good, then *q* is the minimum number of end-numbers. The second of the aims listed above, i.e. to specify the compositions of the hypothetical end-members, is achieved through the factor scores matrix $\mathbf{F}$ ($p \times q$), the factors being the hypothetical end-members. Each row of the factor loadings matrix $\mathbf{H}$ ($N \times q$) gives the contributions of the *q* theoretical end-members to the corresponding object, thus achieving the third of the aims listed above.

If $\mathbf{F}$ has orthonormal columns, so that $\mathbf{F'F} = \mathbf{I}$, then from equations (8.2) and (8.4)

$$\mathbf{C} = \mathbf{HF'FH'} = \mathbf{HH'} \tag{8.5}$$

Matrix $\mathbf{C}$ may be factored

$$\mathbf{C} = \mathbf{ULU'} \tag{8.6}$$

where $\mathbf{U}$ ($N \times r$) has orthonormal columns which are the eigenvectors of $\mathbf{C}$, and $\mathbf{L}$ ($r \times r$) is a diagonal matrix of corresponding eigenvalues. From equations (8.5) and (8.6) the factor loadings matrix is

$$\mathbf{H} = \mathbf{U}\mathbf{L}^{1/2} \tag{8.7}$$

i.e. the eigenvectors scaled to have sums of squares equal to the corresponding eigenvalues. The row sums of squares are then equal to the communalities, which are unity for the complete eigenvector matrix because the elements of the principal diagonal of $\mathbf{C}$ are unity (cf. principal component analysis and R-mode factor analysis). Thus, with q columns in $\mathbf{H}$ a row sum of squares represents the proportion of the total compositional information in that object which is accounted for. In many cases the analysis may end with the factor loadings matrix. However, factor scores are calculated easily (e.g. Jöreskog *et al.* 1976). Premultiplying equation (8.4) by $\mathbf{H}'$ we get

$$\mathbf{H}'\mathbf{W} = \mathbf{H}'\mathbf{H}\mathbf{F}' \tag{8.8}$$

From equation (8.7), and because $\mathbf{U}'\mathbf{U} = \mathbf{I}$

$$\mathbf{H}'\mathbf{H} = \mathbf{L} \tag{8.9}$$

so that

$$\mathbf{H}'\mathbf{W} = \mathbf{L}\mathbf{F}' \tag{8.10}$$

hence

$$\mathbf{F} = \mathbf{W}'\mathbf{H}\mathbf{L}^{-1} \tag{8.11}$$

The singular value decomposition of $\mathbf{W}$ is

$$\mathbf{W} = \mathbf{U}\mathbf{G}\mathbf{V}' \tag{8.12}$$

where $\mathbf{V}$ ($p \times r$) has orthonormal columns which are the eigenvectors of $\mathbf{W}'\mathbf{W}$, and $\mathbf{G}$ ($r \times r$) is a diagonal matrix of singular values which are the positive square roots of the eigenvalues of $\mathbf{W}'\mathbf{W}$ and $\mathbf{W}\mathbf{W}'$. By analogy with equation (2.4)

$$\mathbf{V} = \mathbf{W}'\mathbf{U}\mathbf{G}^{-1} \tag{8.13}$$

From equation (8.7)

$$\mathbf{H} = \mathbf{U}\mathbf{G} \tag{8.14}$$

so that

$$\mathbf{U}\mathbf{G}^{-1} = \mathbf{H}\mathbf{L}^{-1} \tag{8.15}$$

hence

$$\mathbf{V} = \mathbf{W}'\mathbf{H}\mathbf{L}^{-1} = \mathbf{F} \tag{8.16}$$

The required factor scores are the eigenvectors of $\mathbf{W'W}$. Also, post-multiplying both sides of equation (8.4) by $\mathbf{F}$

$$\mathbf{H} = \mathbf{WF} \qquad (8.17)$$

Geometry

Imbrie and Purdy (1962) illustrated the method using a simple hypothetical case in which the system being studied is described completely by three attributes (see also Jöreskog *et al.* 1976). The elements of a given row of matrix $\mathbf{W}$ are the co-ordinates, on rectangular axes representing the attributes, of a point representing the end of the row vector. Because the row vectors are of unit length, those points will lie on the surface of a right spherical triangle of unit radius. Furthermore, if we construct a perpendicular from an attribute axis to the end of a unit row vector we obtain a right-angled triangle with unit hypotenuse. The cosine of the angle between the vector and the attribute axis is equal to the length of the adjacent side, i.e. for the ith row and the jth attribute is equal to w_{ij}. That is, the elements of $\mathbf{W}$ are direction cosines of the row vectors with respect to the attribute axes. The degree of similarity between any two objects is represented by the cosine of the angle between their vectors, ranging from zero (cos 90°) to unity (cos 0°); hence the cosine θ similarity measure. The geometry generalizes to p attributes.

The elements of a given row of $\mathbf{H}$, the factor loadings matrix, are the co-ordinates of the row (object) vector on the orthogonal factor axes. An element h_{ij} is the cosine of the angle between vector $\mathbf{w}'_i$, representing the ith object, and the jth factor axis. Because the row vectors of $\mathbf{W}$ are of unit length it is also the projection of the ith row vector onto the jth factor axis. The sum of squares of a row of matrix $\mathbf{H}$ (i.e. the communality), with all r columns, is unity, so that the square of any element represents the proportion of the compositional information in that object which is accounted for by the corresponding factor. The row sum of squares for $q < r$ factors is a measure of the information in that object which is accounted for by the q factors. The sum of the squared elements in a column of $\mathbf{H}$ represents the sum of projected vector lengths on the corresponding factor axis, and is numerically equal to the eigenvalue. The square of any element $h_{ij}{}^2$ represents the information contributed by the ith object to the jth axis.

Type of data and transformation

Because the concept of the analysis is to identify mixtures of sources of a given composition which will result in the observed compositions of objects, the data should be in the form of relative frequencies, i.e. proportions or percentages. Multiplication of the values for any object or objects by a constant will not affect the result, but other types of transformation will. If the attributes are not homogeneous, e.g. if trace elements are recorded on one scale and major elements on another, the attributes should be transformed to a uniform scale. One such transformation is to record each value as a proportion or percentage of the observed maximum, which has the effect of giving each attribute an equal influence in determining the sample composition as given by the vector position (Imbrie and van Andel 1964).

Computation

1. From the data matrix $\mathbf{X}$ ($N \times p$), form the diagonal matrix $\mathbf{D}$ of row sums of squares and compute $\mathbf{W}$ as in equation (8.3).

2. Compute the similarity matrix $\mathbf{C} = \mathbf{WW}'$.

3. Compute eigenvectors $\mathbf{u}_j$ and eigenvalues l_j of $\mathbf{C}$.

4. Compute the factor loadings matrix $\mathbf{H} = \mathbf{UL}^{1/2}$ i.e. the eigenvectors normalized to have sums of squares equal to the corresponding eigenvalues.

5. If required, rotate matrix $\mathbf{H}$ using one of the standard procedures discussed for R-mode factor analysis.

6. Form the factor scores matrix as in equation (8.11).

Because matrix $\mathbf{C}$ is of order $N \times N$, computer memory may be inadequate to enable its eigenvalues and eigenvectors to be found by usual methods. One solution is to use the TRED3, RATQR, TRISTURM, TRBAK3 sequence suggested for principal co-ordinate analysis; or another memory-saving method. Alternatively, the eigenvectors $\mathbf{v}_j$ and eigenvalues l_j of $\mathbf{W}'\mathbf{W}$, of order $p \times p$, may be found using the method given for principal component analysis. From equation (8.16) we know that $\mathbf{V} = \mathbf{F}$, the factor score matrix, and from equation (8.17) $\mathbf{H} = \mathbf{WV}$. This is the procedure given in Klovan and Imbrie (1971).

Alternatively, if sufficient computer space is available, the singular value decomposition could be computed using the algorithms given by Golub and Reinsch (1971). This has the advantage of eliminating numerical inaccuracy involved in the additional matrix manipulations required above.

Dimensionality, interpretation and rotation

The first of our aims, i.e. to find the minimum number of end-members is achieved by deciding on a value for q, the number of dimensions required to give a good approximation to **W**. The total information on the composition of any object is the sum of the squares of the observations recorded for that object. The transformation in equation (8.3) does not affect the proportions of the constituents, but it gives each row of matrix **W** a unit sum of squares. These unit row sums of squares form the principal diagonal of matrix **C**. The total information for all objects is the trace of **C**, and it is numerically equal to N. The eigenvalues of **C** represent portions of the total compositional information in the row-standardized data that can be accounted for by the factors. Therefore, each eigenvalue expressed as a percentage of N is a convenient expression of the contribution of the corresponding factor (end-member) to the total information, and cumulative percentages can be found for eigenvalues in order of decreasing magnitude.

There is no test for the required number of end-members, and as in many other ordination methods, both R- and Q-mode, the user has to use a rule-of-thumb. One possibility is to decide, arbitrarily, on an acceptable percentage of the total information accounted for. Another method is to plot the eigenvalues against the factor numbers and look for an obvious break in the line.

The sum of squares of a column of **H** equals the corresponding eigenvalue, and the sum of squares of a row represents the total compositional information in that object that is accounted for by all the q theoretical end-members. The row sums of squares are the communalities, which are equal to the principal diagonal elements of **C**, and are therefore unity for r end-members. The square of any element h_{ij}^2 therefore represents (a) the actual contribution of the ith object to the information contained in the jth factor (end-member), and (b) the proportion of the information in the ith object that is accounted for by the jth end-member.

The columns of matrix **F**, which specify the compositions of the hypothetical end-members (factors), have unit sums of squares. Therefore, the proportion of the information in each hypothetical end-member which is contributed by a given attribute is found by squaring the appropriate element of **F**.

In practice, the user might not be interested in theoretical end-members, especially as they are likely to be mathematical abstractions which are difficult to interpret. The solution is to rotate the factor loadings matrix **H** so that individual factor axes coincide with, or are close to, vectors representing observed objects. That is, the aim is to find one object which would have a loading of unity for one of the q factors and zeros for the others. This corresponds to the concept of simple structure in rotation of R-mode factor axes, and one of those rotation procedures can be used. An orthogonal rotation procedure such as varimax may prove useful, and the row sums of squares (communalities) are not altered by orthogonal rotation, although the column sums of squares are. However, Imbrie and van Andel (1964) noted that even after orthogonal rotation, two disadvantages remain:

(a) the composition of the theoretical end-members is controlled to a considerable extent by chance sampling variations ('noise'), and (b) there is no assurance that the composition of these end-members will be easier to interpret after orthogonal rotation. Both disadvantages can be overcome by using an oblique rotation procedure. After rotation, the eigenvalues expressed as a percentage of N may still be used as measures of the information content, or 'explanatory power', of the factors (example in Jöreskog *et al.* 1976).

Davis (1973) drew attention to the fact that because a set of end-members found by oblique rotation will be intercorrelated, the objects selected will not be 'pure', even though they are the most extreme objects in the data matrix. This may not matter if the purpose of the analysis is simply to arrange the objects on a gradient, but if we wish to draw inferences outside the data set it may be better to have 'pure' end-members, even if they are entirely hypothetical. Of course, the factor loadings matrix refers to row-standardized data. In order to recover the composition of each object in terms of unstandardized end-members, the columns of **H** must be divided by the square roots of the object sums of squares (Imbrie and van Andel 1964 p1141). If each element in a row of the resulting matrix is expressed as a proportion of the row sum the resulting matrix will be an oblique compositional matrix. Normally this will not be necessary.

When a factor analysis has been performed on a population, it may be desirable to extend its results to another sample from the same population without analysing the sample separately. For example, Imbrie and Kipp (see Klovan and Imbrie 1971) analysed foraminiferal data from samples widely distributed over the Atlantic sea bottom, and then wished to describe fossil assemblages from one deep-sea core in terms of the varimax factors gained from the general sea-bottom study. This was possible because the same species occurred in both data matrices. In equation (8.17) factor loadings matrix **H** for one core could be obtained from the factor scores matrix **F** from the overall study. In a similar way matrix **F** can be used to find the centroid vector from the row vector of normalized mean values.

The factor loadings matrix **H** gives the co-ordinates, on orthogonal factor axes, of points locating the ends of vectors which represent the objects. Either the points alone may be plotted (e.g. Jöreskog *et al.* 1976) or the vectors may be drawn (e.g. Davis 1973). Because the vectors are of unit length, they will be plotted in two dimensions as radii of a circle. The angles between the radii indicate the interobject similarities. The objects lying closest to the factor axes may be taken as representing the end-members, with a tolerance limit indicated by the communalities (e.g. Imbrie and Purdy 1962).

Such plots can be scrutinized for clusters, and any cluster found can be named by reference to the chief constituents of the objects in it. Indeed, a cluster analysis may be applied to those points (e.g. Imbrie and Purdy 1962) using cos θ as an interobject measure. Imbrie and Purdy also described a method for overcoming the problem of clustering with a large number of objects. However, Davis (1973) pointed out that if the objective is to search for groups or clusters, then either a direct cluster analysis or an R-mode analysis may provide a computationally more efficient means for doing so. He demonstrated that a dendrogram constructed by weighted pair-group averaging using the correlation matrix gave exactly the same relative arrangement of the objects as did a Q-mode factor analysis. However, if the attributes sum to a constant value (ipsative measures), the R-mode correlation matrix may contain spurious negative correlations which may make the analysis useless. In such a case a Q-mode analysis, which is carried out across attributes, will be meaningful (Davis 1973).

Applications of Q-mode factor analysis

Imbrie and van Andel (1964) described two case-studies representing different, but common, situations in sedimentary petrology. In the Gulf of California, most sedimentary mineral assemblages are derived from nearby, petrographically simple, sources and are dominated by only a few minerals. There is little mixing during transportation, and the system can be defined easily in terms of a few mineralogically-distinct end-members. Q-mode factor analysis of this system, with oblique rotation, gave results similar to those obtained by a conventional inspection of the data. On the other hand, the Orinoco-Guayana Shelf has remote and petrographically-complex sources, and mixing during the long-distance transportation is common. The mineral assemblages are complex and variable, and obvious end-members are lacking. Q-mode factor analysis of this system, with oblique rotation, gave a mineral distribution pattern greatly different from that obtained by inspection of the data, and the Q-mode result appeared to be more meaningful when interpreted in terms of zones of littoral transportation moving landward during the post-Pleistocene rise of sea level.

Imbrie and Purdy (1962) applied the method to the classification of Bahamian carbonate sediments. Twelve attributes, expressed as weight per cent or volume per cent, were measured on 40 samples (objects). In order to give all attributes equal weight in determining the location of any object vector, each attribute in the raw data matrix was expressed as per cent of the maximum value observed. If this is not done, attributes contributing a large fraction of the total volume would tend to exert a large influence on the position of an object vector. Of course, this may be acceptable in some studies. After orthogonal (quartimax) rotation, the positions of points representing the objects were plotted on pairs of factor axes using the co-ordinates given in the factor loadings matrix. It was clear that the 40 objects fell into five discrete groups which were named according to the chief constituents of the objects in the groups. The four objects closest to the factor axes were taken to be the end-members. Within a tolerance imposed by the communalities, each of the 40 objects could be represented as a simple mixture of these end-members. Imbrie and Purdy also described a hierarchical cluster analysis method in which each cluster was represented by a centroid vector of unit length. The cosines of the angles between all pairs of cluster centroid vectors were then calculated and the pair with the largest cosine (i.e. greatest similarity) were joined together and a new group centroid vector was obtained.

Martini and Acton (1975) used the method of Klovan and Imbrie (1971), with oblique rotation, on 23 attributes (11 chemical, 12 particle size classes) measured on 120 samples collected from five lacustrine soils. Three factors, interpreted as indicators of depositional and weathering environments, accounted for 97% of the variation in the data. When the chemical and particle size data were treated separately, two factors, interpreted as representing weathering processes in the soil profile, accounted for 97% of the variation in the chemical data. Three factors, interpreted as representing energy levels in the deposition environment of the parent materials, accounted for 99% of the variation in the particle size data.

Dean and Gorham (1976) found that Q-mode factor analysis, with varimax rotation, of 43 chemical and mineralogical attributes in profundal sediments from 46 Minnesota lakes permitted subdivisions of the water-chemistry groups based mainly on organic- and carbonate-related attributes.

Advantages and disadvantages of Q-mode factor analysis

Obviously, if it is of interest to study the relative compositions of objects in relation to hypothetical or real end-members, then it is advantageous to use Q-mode factor analysis. As the method is descriptive, rather than statistical, there are very few requirements to be complied with. The attributes must be of a type which can be added or multiplied, usually proportions or percentages. The method is easy to apply and to interpret. It is particularly advantageous if the attributes sum to constant values (ipsative measures), because such attributes give rise to spurious correlations which may make an R-mode analysis useless. A Q-mode analysis avoids this problem.

CHAPTER 9

CORRESPONDENCE ANALYSIS

Correspondence analysis is an essentially graphical technique for displaying the rows and columns of a two-way contingency table as points in corresponding low-dimensional Euclidean vector spaces. It is an extension of an older technique, and it involves geometric ideas similar to Karl Pearson's original view of principal component analysis in which the closeness of points to a low-dimensional subspace is given by the sum of the squared distances from the points to the subspace. Correspondence analysis was developed by Benzécri (1973), and in geometrical terms involves defining firstly a cloud of points in a multidimensional vector space, secondly the metric structure on this space, and thirdly the fit of the cloud of points to a low-dimensional subspace onto which the points are projected for display and interpretation. There are, of course, two clouds of points, one representing the rows and one representing the columns of the matrix, and the displays of the points in the low-dimensional vector spaces may be superimposed to give a joint display. That is to say, the aim is to obtain, simultaneously and on equivalent scales R-mode and Q-mode loadings. One of the main advantages of the method is that it allows the simultaneous portrayal of graphical relationships between objects and attributes. Benzécri's geometrical ideas involve the concepts of mass, inertia, chi-square distance, profile, and centre of gravity, which are discussed in detail below.

The geometrical approach to correspondence analysis broadens its base, so that with suitable care it may be extended to a range of matrices of non-negative data. The method is theoretically equivalent to a number of techniques, many of which have been in use for some time. It may be described as a special case of canonical correlation analysis (Hill 1974) and a form of discriminant analysis. A very readable account is given in Greenacre (1981) and a less-detailed account is given in Jöreskog *et al.* (1976). A detailed and highly-mathematical account is given in Greenacre (1978), who showed that correspondence analysis belongs to a class of methods which includes the biplot (Gabriel 1971, 1981) and principal component analysis, which are all linked by the notion of the basic structure of a matrix. Probably the best account in English is that of Greenacre (1984), who also discussed the related methods reciprocal averaging,

129

Table 9.1 Frequencies of three antelope tribes in five African game reserves (Greenacre 1981)

	Game Reserve	Tribe A	B	C	Row sum
A. Recorded frequencies	1	153000	11124	0	164124
	2	8000	5450	0	13450
	3	0	6569	24041	30610
	4	65000	9500	190000	264500
	5	0	400	5000	5400
	Sum	226000	33043	219041	478084
B. Row profiles	1	0.932	0.068	0	1
	2	0.595	0.405	0	1
	3	0	0.215	0.785	1
	4	0.246	0.036	0.718	1
	5	0	0.074	0.926	1
					(mass)
C. Proportions	1	0.320	0.023	0	0.343
	2	0.017	0.011	0	0.028
	3	0	0.014	0.050	0.064
	4	0.136	0.020	0.397	0.553
	5	0	0.001	0.011	0.012
	Sum (mass)	0.473	0.069	0.458	1.000
D. Column profiles	1	0.677	0.337	0	
	2	0.035	0.165	0	
	3	0	0.199	0.110	
	4	0.288	0.287	0.867	
	5	0	0.012	0.023	
	Sum	1	1	1	

dual scaling, canonical correlation analysis of contingency tables, and simultaneous linear regressions.

Theory

Table 9.1A gives a 5 × 3 matrix of frequencies of three antelope tribes in five African game reserves (Greenacre 1981). This table may be re-expressed in two ways. Firstly, we can divide each element of a row by the row sum, so that the new row sums are all unity (Table 9.1B). Here, the row elements are relative frequencies for each reserve (object), and a row vector is called a 'row profile'. Column profiles can be obtained in a similar way. Secondly, we can divide each element (frequency) of Table 9.1A by the grand sum, as in Table 9.1C. Here, all the elements in the table sum to unity, and they are proportions that may be interpreted as probability values, so that Table 9.1C is a contingency table in the sense of Fisher (1940). The vector of row sums contains the marginal probabilities for the objects, the vector of column sums contains the marginal probabilities for the attributes.

Also, in Table 9.1C, the row (or column) sums are effectively masses (or weights) which sum to unity. This is evident from the fact that if we multiply the elements of a row profile by the corresponding mass, we get the appropriate row of Table 9.1C. The centre of gravity of the row profiles (i.e. the average profile) is the sum of the masses times the profiles, i.e. the vector of column sums in Table 9.1C. This is so because the centroid of N points $\mathbf{x}_i$ with different weightings w_i is the weighted average point

$$\bar{\mathbf{x}} = \sum_{i=1}^{N} w_i \mathbf{x}_i \bigg/ \sum_{i=1}^{N} w_i$$

Similarly, if $\mathbf{a}_i'$ is a vector of relative frequencies (row profile) with weightings n_i (observed row sums) then

$$\bar{\mathbf{a}} = \sum_{i=1}^{N} n_i \mathbf{a}_i' \bigg/ \sum_{i=1}^{N} n_i$$

For any row i

$$n_i \bigg/ \sum_{i=1}^{N} n_i$$

is the corresponding mass (row sums in Table 9.1C). Hence the centroid $\bar{\mathbf{a}}$ is the sum of the masses times the row profiles.

It is clear from the above that the profile of a given object (reserve) is independent of the number of items (animals) sampled, i.e. the row total, but the latter is a measure of the importance of the profile in the analysis. This concept of relative masses was central to Benzécri's ideas about correspondence analysis. In principal component analysis, however, all points effectively have equal mass.

The next important idea in correspondence analysis is the development of a suitable metric for expressing the distances between the rows and between the columns, because Pythagorean distance is not applicable to contingency data of the type used here. In order to produce both R-mode and Q-mode loadings in the same metric, the latter must be defined jointly and symmetrically. If we let $\mathbf{P}$ ($N \times p$) be a matrix of proportions, as in Table 9.1C, then $\mathbf{P}$ is transformed to $\mathbf{Q}$ ($N \times p$) by

$$\mathbf{Q} = \mathbf{R}^{-1/2}\mathbf{P}\mathbf{C}^{-1/2} \tag{9.1}$$

where $\mathbf{R}$ and $\mathbf{C}$ are diagonal matrices of row and column sums of $\mathbf{P}$ respectively. The effect of this transformation is to stretch the column vectors by the reciprocals of the square roots of the corresponding column sums and the row vectors by the reciprocals of the square roots of the corresponding row sums. It is usual, but not essential, for each element in matrix $\mathbf{P}$ to be doubly centred on its row and column, i.e.

$$q_{ij} = \frac{p_{ij} - r_{ii} \times c_{jj}}{\sqrt{r_{ii} \times c_{jj}}} \tag{9.2}$$

This centring step eliminates in advance the largest eigenvalue, which is equal to unity in correspondence analysis and has no meaning.

The rationale for the transformation in equations (9.1) and (9.2) is connected with the definition of a chi-square distance metric appropriate to the rows and columns. The distance between rows i and j is given by

$$d_{ij} = \sum_{k=1}^{p} \frac{1}{c_{kk}} \left[\frac{p_{ik}}{r_{ii}} - \frac{p_{jk}}{r_{jj}} \right] \tag{9.3}$$

(e.g. see Jöreskog *et al.* 1976; Legendre and Legendre 1983).

This is equivalent to

$$d_{ij} = (\mathbf{a}_i - \mathbf{a}_j)' \, \mathbf{C}^{-1} (\mathbf{a}_i - \mathbf{a}_j) \tag{9.4}$$

where $\mathbf{a}_i$ is the ith row profile. With this distance measure, the larger differences between attributes with large frequencies are reduced and the smaller differences between attributes with low frequencies are increased, so that the contributions of rare and frequent objects tend to be equalized (Greenacre 1981).

At this stage, we have the row profiles represented by points with different masses in a chi-square metric space. The next step is to find a lower-dimensional subspace which is closest to all the points. To do this we need a simple cross-products matrix from which to extract principal components. We use the modified covariance matrix

$$\mathbf{W} = \mathbf{Q}'\mathbf{Q} \tag{9.5}$$

(e.g. see Jöreskog *et al.* 1976). The eigenvalues λ_j and eigenvectors $\mathbf{u}_j$ of $\mathbf{W}$ ($p \times p$) are found. The eigenvectors are the principal axes of inertia, and are orthonormal in the metric $\mathbf{C}^{-1}$. The eigenvalues, in descending order, are the moments of inertia along the principal axes. The trace of $\mathbf{W}$ is equal to the total inertia, which is also equal to the mean square contingency coefficient calculated from the original data matrix (Greenacre 1984 p86).

So far, we have discussed the projection of the cloud of row profiles onto a low-dimensional subspace, that is, an R-mode analysis. Everything that has been said about the row profiles applies to the Q-mode analysis of the column profiles. We can find the eigenvalues λ_j and eigenvectors $\mathbf{v}_j$ of $\mathbf{Q}\mathbf{Q}'$ ($N \times N$). The eigenvalues will be the same as those obtained above because the eigenvalues of a minor product moment matrix are the same as those of the corresponding major product moment matrix, and the eigenvectors are the principal axes of inertia which are orthonormal in the metric $\mathbf{R}^{-1}$.

In fact, the eigenvalues and both sets of eigenvectors are conveniently obtained from the singular value decomposition of matrix $\mathbf{Q}$

$$\mathbf{Q} = \mathbf{V}\mathbf{G}\mathbf{U}' \tag{9.6}$$

where $\mathbf{G}$ ($r \times r$) is a diagonal matrix of singular values, which are the positive square roots of the eigenvalues. $\mathbf{V}$ is of order $(N \times r)$ and $\mathbf{U}$ is of order $(p \times r)$.

Having obtained the two sets of eigenvectors, the final step is to express them in a common metric, thus obtaining matrices $\mathbf{X}$ $(N \times r)$ and $\mathbf{Y}$ $(p \times r)$, which contain respectively the co-ordinates of the rows and columns on common rectangular axes

$$\mathbf{X} = \mathbf{R}^{-1/2}\mathbf{VG} \tag{9.7}$$

$$\mathbf{Y} = \mathbf{C}^{-1/2}\mathbf{UG} \tag{9.8}$$

(Greenacre 1978 p 56). For a given principal axis, the relationships between the co-ordinates $\mathbf{X}$ for the row points and $\mathbf{Y}$ for the column points are given by the 'transition formulae'

$$\mathbf{X} = \mathbf{AYG}^{-1} \tag{9.9}$$

and

$$\mathbf{Y} = \mathbf{B'XG}^{-1} \tag{9.10}$$

where $\mathbf{A}$ is the matrix of row profiles as in Table 9.1B, $\mathbf{B}$ is the matrix of column profiles as in Table 9.1D.

Thus there is a duality and symmetry in the formulations and solutions of these dimension-reducing problems, and a correspondence of the row and column points on the new axes which account for the name correspondence analysis.

Type of data

Correspondence analysis is intended primarily for the analysis of contingency tables. The rows may represent objects and the columns attributes, but the contingency table could compare the various states of two sets of attributes, e.g. species and environmental factors. Care is necessary when deciding whether or not the analysis is applicable to other types of data table. Greenacre (1978) posed some questions which indicate whether or not this method should be applied:

1. Is the strategy of finding the principal axes of inertia meaningful in the context of the study in question, i.e. is it reasonable to think in terms of profiles, masses, and chi-square distances for at least one of the clouds of points?

2. How important is the simultaneous display of the rows and columns, and are the transition formulae sufficient justification for the between-sets positions?

From the theory given above, it is obvious that the elements of the data table must be in the same units so that they can be added, and must be non-negative so that they can be transformed into probabilities. Some data, such as species abundance tables, already have these properties. Other data may need recoding. Apart from frequency tables of the type already discussed, correspondence analysis is applicable to tables containing binary data, data scored on an arbitrary scale (e.g. 0 to 10), and mixed data types. A matrix of distances or similarities is not suitable for correspondence analysis, instead Greenacre (1978) recommended an unfolding analysis and gave an algorithm for its application.

Greenacre (1981) gave some examples of the application of the method to different types of data. For arbitrary scale scores it may be useful to add a new column containing the complementary score alongside each existing column. This is called 'doubling' and it has the effect of making the row sums equal, thus giving all rows the same mass. Quantitative data could be converted to qualitative data by dividing the range of each quantitative attribute into a small number of intervals and scoring each value in an interval as 1 or 0 on the corresponding new binary attribute. A contingency table is then easily obtained. Such a conversion needs to be carried out with great care, as it involves a loss of information. Jambu and Lebeaux (1983) suggested that if histograms of the attributes do not show evident peaks and troughs, then division into classes with equal numbers of objects is required although this option can have the undesirable result of objects being put into a class for reasons other than their similarity (see Section 3.1). Greenacre (1981) suggested a method for recoding which reduces the information loss.

The fact that, after recoding, each object contributes to several of the table's frequencies does not affect the results because of the transformations involved in the method. In fact, the rows of the table could be used for different species, each species being divided into a few abundance classes. The columns could be used for attributes of the physical environment (Legendre and Legendre 1983).

Greenacre (1984 pp 76–80) discussed the analysis of data having attributes of constant sum (ipsative data), such as are often obtained in soil science and geology.

A two-way contingency table is the condensation of a data matrix in which the entries are either 1 or 0. Such a matrix is called an indicator matrix, an incidence matrix, or a response-pattern matrix. The columns are called variously dummy variables, pseudo-variables, or indicator variables. In the simplest form of indicator matrix, the entries indicate to which of the discrete attribute categories each object (row) belongs. In another form of indicator matrix, the rows may represent another set of attribute categories. However, if an additional discrete attribute has also been recorded, the analysis of the indicator matrix and any two-way frequency table derived from it might differ quite considerably from that of a bivariate table. The most common example of multivariate indicator matrix arises in questionnaires. The analysis of such a matrix is called multiple correspondence analysis, and is rather complicated. Interested readers should consult Greenacre (1984).

Computation

1. Begin with a suitable data table (e.g. Table 9.1A).

2. Compute matrix **Q** as in equation (9.1) or (9.2). Note that if the data are not centred, i.e. if equation (9.1) is used, the first eigenvalue will be unity, and it and the corresponding principal axis must be ignored.

3. Compute the singular values and the eigenvector matrices **V** and **U** by the singular value decomposition of **Q**. Alternatively, if computer memory space is limiting, compute the eigenvalues and eigenvectors of **QQ′** or **Q′Q**, whichever has the smaller order, using the method given for principal component analysis.

4. Compute matrices **X** and **Y** as in equations (9.7) and (9.8). Alternatively, if the eigenvalues and eigenvectors of **QQ′** or **Q′Q** have been obtained, compute matrix **X** or **Y** as appropriate, and compute the required matrix using the appropriate transition formula in equation (9.9) or (9.10).

The singular value decomposition (equation 9.6) is the preferred method, as it eliminates numerical inaccuracy involved in the other matrix manipulations. A suitable algorithm is given in Golub and Reinsch (1971).

Rows and columns of new points may be represented on an existing display by applying the appropriate transition formula (Greenacre 1984 p71). Because such rows and columns do not contribute to the chi-square distance or to the determination of the principal axes, their contribution to the axes is not defined and it is convenient to think of them as having zero mass.

Reciprocal averaging

In the direct gradient analysis of plant species data, some well-marked environmental gradient is selected and scores are assigned to species according to their suspected preferences for position on the gradient. Sites may then be ordinated by taking averages of the scores of the species which occur in them. However, the physical basis for a floristic gradient is commonly unknown, in fact it may not even be obvious that there is a gradient. In such a case, an indirect gradient analysis is performed using an ordination method. Principal component analysis does not give acceptable results if the species show markedly non-linear distributions along the gradient. In an attempt to overcome this problem, Hill (1973) used a form of correspondence analysis which he called reciprocal averaging.

The method begins with a presence–absence (binary) data matrix, the rows representing species and the columns representing stands (or vice-versa, the method is symmetrical with respect to rows and columns). An initial gradient analysis is obtained by allocating to each species a score, from 0 to 100, according to its known or suspected position on the gradient of interest. This gives a vector $\mathbf{x}_0$. Initial stand scores are obtained by averaging the initial scores for the species in each stand. This gives a vector $\mathbf{y}_0$. A new vector of species scores $\mathbf{x}_1$ is obtained by averaging the scores of the stands in which the species occur. Then a new vector $\mathbf{y}_1$ of stand scores is calculated, and the process is iterated until the scores stabilize. The final scores do not depend on the initial scores, although the number of iterations required does.

The algebra of this method (Hill 1973, 1974; Mardia *et al.* 1979) is as follows. If $\mathbf{R}$ and $\mathbf{C}$ are diagonal matrices of row and column sums respectively of the incidence matrix $\mathbf{A}$, then

$$\mathbf{x} \propto \mathbf{R}^{-1}\mathbf{A}\mathbf{y} \tag{9.11}$$

$$\mathbf{y} \propto \mathbf{C}^{-1}\mathbf{A}'\mathbf{x} \tag{9.12}$$

Hence

$$\mathbf{x}_1 \propto \mathbf{R}^{-1}\mathbf{A}\mathbf{C}^{-1}\mathbf{A}'\mathbf{x}_0 \tag{9.13}$$

$$\mathbf{y}_1 \propto \mathbf{C}^{-1}\mathbf{A}'\mathbf{R}^{-1}\mathbf{A}\mathbf{y}_0 \tag{9.14}$$

so that $\mathbf{x}$ is an eigenvector of $\mathbf{R}^{-1}\mathbf{A}\mathbf{C}^{-1}\mathbf{A}'$ and $\mathbf{y}$ is an eigenvector of $\mathbf{C}^{-1}\mathbf{A}'\mathbf{R}^{-1}\mathbf{A}$. If the proportionality coefficient is unity, the largest eigenvalue will also be unity. Hill (1973) showed that equation (9.13) is easily transformed to the major product moment form of equation (9.5).

An alternative way of obtaining vectors $\mathbf{x}$ and $\mathbf{y}$ is to let the species score equal the average stand score and the stand score equal the average species score, in both cases rescaling so that the total range is 0 to 100. Solutions thus obtained are the analogues of axes other than the first axis in a principal component analysis. Thus, reciprocal averaging produces a stand ordination comparable to one produced by principal component analysis, plus a species ordination which is more meaningful than are the eigenvector weightings produced by principal component analysis. However, the main advantage of reciprocal averaging is that a good species ordination is used to derive the stand ordination and vice versa. If two species have similar distributions but one is rarer, they will both have much the same species score in reciprocal averaging, but in an unstandardized principal component analysis the rarer species will usually have a smaller axis score. If the data are standardized so that each species has unit variance, principal component analysis may give an ordination closely resembling that obtained by reciprocal averaging of the unstandardized data (Hill 1973).

In reciprocal averaging, as in principal component analysis, the second axis is independent of the first only when the data have a multivariate normal distribution, which is not likely with presence–absence data. Thus, a plot of the points on the first two axes often has a curved or inverted V shape in both species and stand ordinations.

As species presence–absence matrices are usually sparse, a great economy of computer storage can be obtained by using a direct iteration algorithm, Hill (1973) gave details. Lebart (1982) discussed the problem of handling large sparse matrices in the analysis of textual data.

Ter Braak (1983, 1987) described an extension of reciprocal averaging which he called canonical correspondence analysis. In this method, ordination axes are chosen with the restriction that they must be linear combinations of known environmental attributes, which may be quantitative or nominal. As many axes can be extracted as there are environmental attributes. Detrending can be incorporated to remove 'arch' effects. The method is said to be effective when species have bell-shaped response curves or surfaces with respect to environmental gradients. The ordination diagrams produced show the patterns of variation in community composition that can be explained best by the environmental attributes, as well as the main features of species distributions. With test data the ordination diagrams effectively summarized the relationships between environmental attributes and communities of hunting spiders, dyke vegetation, and algae.

Dual scaling

This is another variant of correspondence analysis. In this approach, numerical values are assigned to categories (columns) so as to discriminate optimally, in the least-squares sense, between objects (rows). After scaling, the between-row variance is maximized with respect to the within-row variance. In practice, it is the ratio within-row/total sum of squares that is minimized and the ratio between-row/total sum of squares that is maximized. The latter is called the squared correlation ratio, as it is also the squared canonical correlation between the rows and columns of the data matrix.

Dimensionality

The total dimensionality of a set of row and column profiles is always one less than the minimum of the number of rows and the number of columns. The quality of the display obtained by plotting points representing the rows or columns in the subspace of the first few principal axes is given by the moments of inertia expressed as percentages of the total inertia, i.e. the individual eigenvalues expressed as percentages of the sum of the eigenvalues. The question of how many axes are important involves the question of whether or not a given axis reflects a significant or interesting arrangement of the points rather than residual variation. Approximate significance tests for successive axes for the restricted cases of two-way contingency data and rank-order data have been published (Lebart and Fénelon 1971; Lebart *et al.* 1977), but for more general use a pragmatic guideline is to inspect a graphical display of the percentages of inertia to see if there is an obvious break (Greenacre 1981).

Interpretation

The most important output of a correspondence analysis is the graphical display of points representing the rows and columns. The principal axes tend to reflect the directions of greatest dispersion of a cloud of points, but their orientations are influenced by the masses assigned to the points. In interpreting a scatterplot, it is useful to think of the points being joined to the centre of gravity by lines (vectors) which represent the strengths and directions of the pulls which differentiate the individual points from the centre of gravity.

Probably the first use of such a display will be an attempt to assign a meaning to the axes. As a guide to interpretation, a descriptive name can be given to an axis from the positions of points on the axis and a knowledge of which points have contributed most to the inertia in that axis. The subspace generated by the first q principal axes maximizes the inertia of the cloud of points projected onto it. The sum of the moments of inertia along all the principal axes is the total inertia of the cloud of points, and it is identical to Pearson's mean-square contingency coefficient calculated on the contingency table, i.e. the chi-square statistic divided by the total of the matrix (Greenacre 1981, 1983). This quantity is used as a measure of the total variation within the data matrix, and is decomposed in various ways for interpretation. The total inertia may be decomposed into interpretable inertia and error or random inertia. The interpretable inertia can be further decomposed (Greenacre 1978, 1984).

If r_i is the mass of the ith point, representing a row profile, and x_{ik} is the projection of the point on to the kth axis, i.e. its co-ordinate value which is also its distance along that axis from the centre of gravity, then the inertia of the projection of the ith point is given by $r_i x_{ik}^2$. In general, the principal axis tends towards the points of greater mass, but there are cases where a point with small mass may have a large contribution to the total inertia because of its large distance from the centroid.

As the moment of inertia of the points along an axis is given by the eigenvalue,

$$\sum_{i=1}^{N} r_i x_{ik}^2 = \lambda_k$$

The chi-square distance d_i^2 of the ith profile from the centroid is calculated as the weighted sum of the squared differences between the profile and the average profile, the weights being the inverses of the average profile elements (Greenacre 1984 p70). The inertia of the corresponding profile point vector is $r_i d_i^2$ and it can be decomposed along the principal axes. The part of this inertia along the kth axis is $r_i x_{ik}^2$. Expressed as a proportion of the point's total inertia, this is

$$r_i x_{ik}^2 / r_i d_i^2 = (x_{ik}/d_i)^2 = \cos^2 \theta$$

where θ is the angle subtended with the axis by the line joining the point and the centroid.

$\cos^2 \theta$ is a convenient measure of the closeness of each point to the subspace defined by the axes, because the squared cosines for the full set of axes sum to unity. For the first q axes the sum is called the quality of the display of the point. $\cos^2 \theta$ also gives the relative contribution of an axis to the inertia of a point because it is independent of the point's mass. It is also a measure of the extent to which a profile vector 'correlates' with an axis, so that $\cos^2 \theta$ is also referred to as a squared correlation. Generally, a large contribution of a point to the inertia of an axis implies a large relative contribution of the axis to the inertia of the point, but not vice versa.

For each point, representing a row profile, it is useful to list the following:
(i)   Its mass. The total mass of each cloud of points is unity.
(ii)   Its inertia in the total space. This is the sum of the inertias of the projections of the point on the p axes expressed as a proportion of the total inertia, i.e. for the ith point

$$\sum_{k=1}^{p} r_i x_{ik}^2 \left/ \sum_{k=1}^{p} \lambda_k \right.$$

(iii)   The 'quality' of each point's representation in the q dimensional subspace, i.e.

$$\sum_{k=1}^{q} \cos^2 \theta$$

In addition, for each point and each principal axis, we could list the following:

(a)   The co-ordinate of the point on the axis, i.e. x_{ik};

(b)   The relative contribution of the axis to the inertia of the point, that is $\cos^2 \theta$;

(c)   The contribution of the point to the inertia along the axis, i.e. $r_i x_{ik}^2 / \lambda_k$.

Corresponding values can be listed for each point representing a column profile.

Some examples of interpretation are given in Greenacre (1978, 1981). There are advantages and disadvantages of the simultaneous display of the two clouds of points representing the row and column profiles. One advantage lies in the number of different features of the data which can be expressed in a single figure. For each cloud, the display indicates the nature of similarities and dispersion within it. The joint display indicates the correspondence between the clouds, but it is important to note that we cannot interpret distances between points belonging to different clouds, as no such distances have been defined (Greenacre 1984 p65). However, we can look at angles between vectors and axes in a joint display of points representing rows and columns of the contingency table (see Legendre and Legendre 1983 p306).

Furthermore, the correspondence analysis display is not the optimal representation (in a least squares sense) of the within-set distances in a subspace of fixed dimension. The true chi-square distances could become quite distorted in a principal axis display if the masses were widely different (see Greenacre 1978). This has an important consequence if the user wishes to perform a cluster analysis on the co-ordinate values of the points. Greenacre (1984) presented a hierarchical method which is similar to single linkage clustering on the squared chi-square distances between the profiles, except that these distances are weighted by the masses of each pair of clusters being merged.

Legendre and Legendre (1983) pointed out that the quantity $R = \sqrt{\lambda}$ gives the correlation between the ordinations of the rows and columns. The value $(1 - \lambda_j)$ measures the difficulty of ordering, on the *j*th principal axis, the states of the rows of the table from an ordination of the columns, and vice versa (see also Orloci 1978). Williams (1952) gave methods for testing the significance of $R^2 = \lambda$.

Applications of correspondence analysis

The rationale of correspondence analysis is different from the computationally equivalent technique of reciprocal averaging. The former has a geometric framework, and the basic idea is that of two sets of profile points (or vectors of relative frequencies) with associated masses in multidimensional space. The analysis produces a projection of these points on axes of maximum inertia in a low-dimensional space. In reciprocal averaging, the emphasis is on assigning a set of scores to the rows and columns of a matrix so that the two sets have linear regressions on each other, resulting in an ordination of the rows and columns. Here, the idea of mass is not explicitly included and the analysis seldom proceeds beyond one dimension (Greenacre and Vrba 1984).

Papers on correspondence analysis published in French appeared in the early 1960s and culminated in Benzécri (1973). More recent papers by Bauzon *et al.* (1974), Roux and Salanon (1974), Bonnet *et al.* (1975), Waksman *et al.* (1975), Ménard and Bélanger (1976) and Bonin and Roux (1978) cover a range of topics and illustrate the usefulness of the method. The first English language description of the method was in Benzécri (1969), but there seem to be few examples in English in which this method was used. Teil (1975) discussed geological applications and David *et al.* (1974) found the method to be useful in extracting a maximum of information from chemical analyses of geological samples. Greenacre and Degos (1977) discussed its use in population genetics, and Serre (1977) used it to analyse tree-ring widths as a function of time. Greenacre and Vrba (1984) gave a detailed account of the application of correspondence analysis to antelope census data in African wildlife areas. They found that in their analysis the larger wildlife areas were receiving too great a mass, and therefore different masses were used. They noted that it is not uncommon in correspondence analysis to reweight the points according to additional information, and often this does not radically change the final results.

Gordon (1982) gave examples of the application of correspondence analysis to the representation of stratigraphical levels and pollen taxa of a pollen diagram.

Advantages and disadvantages of correspondence analysis

An important advantage of correspondence analysis is that it is distribution-free and can be used on binary and multistate attributes. Other advantages are that it provides, simultaneously, ordinations of both the rows and columns of the input matrix, and it minimizes the effects of rarity and abundance.

Its chief disadvantage is that it is not well suited for use with continuous attributes. Although methods for recoding continuous attributes have been suggested, this inevitably involves some loss of information and if done badly will distort relationships. Also, although simultaneous graphical representation of the rows and columns of the matrix is obtained, between-sets interpretation of interpoint distances is dangerous because of the incompatibility of the transition formulae with this aim, and for some data the within-set distances obtained may represent the actual distances only weakly.

Reciprocal averaging produces stand ordinations which are very similar

Table 9.2 Counts of fruiting bodies of nine species of fungi on soils in stands of lodgepole pine in four age classes

Fungal species	Age class of lodgepole pine			
	I	II	III	IV
1. Lactarius rufus	805	149	85	3
2. Inocybe longicystis	42	117	45	13
3. Russula emetica	3	22	32	13
4. Clitocybe sp.	6	40	8	280
5. Nolanea cetrata	4	13	0	39
6. Cortinarius croceofolius	24	54	14	6
7. Laccaria sp.	0	48	201	0
8. Galerina sp.	28	134	15	27
9. Paxillius involutus	15	0	0	0

to those from principal component analysis of between-species correlation matrices, i.e. with data centred and standardized. However, if there are many zeros in the presence–absence matrix reciprocal averaging is better than principal component analysis because the chi-square metric and distance are asymmetric coefficients, i.e. they exclude joint absences. This does not happen with Pythagorean distance which is preserved in principal component analysis.

Many authors have used reciprocal averaging on vegetation data and found the results to be satisfactory (e.g. Golden 1979; Kent and Wathern 1980; Cooper 1984). However, Hill and Gauch (1980) drew attention to two faults in the method. One is the 'arch' or 'horseshoe' effect, which is a mathematical artifact not corresponding to any real structure in the data. It arises when the second axis is not independent of the first, and results in the compression of interobject distances at the ends of the first axis (Kershaw and Looney 1985). The other main fault is that reciprocal averaging does not preserve ecological distances. Hill and Gauch (1980) described an extension to reciprocal averaging which they called detrended correspondence analysis. Tests with simulated and field data showed that the above faults were corrected by detrended correspondence analysis, but this method has limitations which make it necessary for the user to remove extreme outliers and discontinuities prior to analysis. However, Fewster and Orloci (1983) commented that detrended correspondence analysis removes from the data trends that the user of non-linear ordinations hopes to detect, and although this may be justifiable and may greatly assist the scrutiny of the ordination, it cannot be condoned as a general strategy.

It should also be noted that Gauch and Wentworth (1976) found that although reciprocal averaging gave good results with vegetation data, results with environmental data were usually poor.

Table 9.3 Eigenvalues (inertias) of the axes obtained from the data in Table 9.2

Eigenvalue	Percentage of inertia	
	Axis	Cumulative
0.6829	52.70	52.70
0.4408	34.01	86.71
0.1723	13.29	100.00

A practical example of correspondence analysis

Table 9.2 gives counts of fruiting bodies of nine species of fungi growing in soil under stands of lodgepole pine (*Pinus contorta* Douglas ex Loudon) belonging to four age classes (Dr. J. Dighton, personal communication). There are three eigenvalues (Table 9.3). Joint displays of the row and column points in the three dimensions are given in Figures 9.1 and 9.2. To interpret these displays we need to consider the decompositions of inertia.

The inertia along the first axis equals the weighted sum of the squared distances from the origin of the row (or column) profiles which are represented by the points, the weights being the masses of the respective points. Hence, we can examine the contributions of the points to the inertias of successive axes, which are given in Table 9.4 as proportions (i.e. the columns sum to unity). Fungal species 4 contributes nearly 67% to the first axis, and species 1 contributes another 24%. If we think of the points as exerting forces of attraction for the axis by virtue of their positions and masses, these two fungal species are largely responsible for the orientation of the first axis. Similarly, lodgepole pine age class IV contributes 75% to the inertia in the first axis and age class I contributes 23%. Figure 9.1 shows that the point representing species 4 coincides with the point representing stand IV, and the point representing species 1 is close to the point representing age class I. However, the first axis is not a simple age sequence axis, as age classes II and III are not ordered by it. It seems to be more an axis which separates the oldest lodgepole pine stand, which appears to be quite different from the rest.

Having interpreted the first axis, we can examine how close to it the points lie. For this we need the relative contributions of the axes to the inertias of the points, which are also the squared cosines of the angles between the point vectors and the axes (squared correlations) given in Table 9.5. The values for the first axis show that 95% of the inertia of species 4 and 61% of that of species 1 are contributed by the first axis. Similarly, nearly 96% of the inertia of lodgepole pine age class IV and 52% of the inertia of age class I are contributed by the first axis. Generally, a large contribution of a point to the inertia of an axis implies a large relative contribution of the axis to the inertia of the point, but not conversely. For

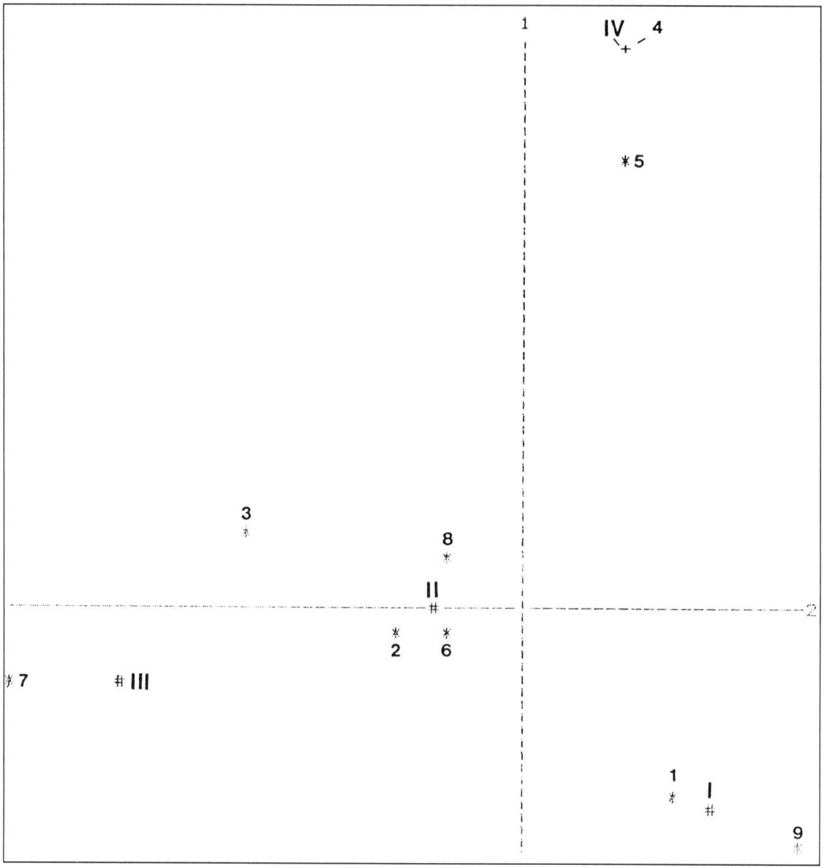

Figure 9.1 Scatterplot of points representing the fungi (rows) and age classes of lodgepole pine (columns) in Table 9.2 on the first two correspondence analysis axes

example, species 5 is highly correlated with the first axis although it contributes very little to the inertia in that axis.

Turning to the second axis, fungal species 7 makes the largest contribution to the inertia and species 1 the next largest (63% and 22% respectively). Lodgepole pine age class III contributes 62% and age class I contributes 30% (Table 9.4). Figure 9.1 shows that species 1 is associated with age class I and species 7 is associated with age class III, and the second axis discriminates between these two combinations. The second axis contributes almost 89% to the inertia of fungal species 7 and 35% to species 1. Other species which are highly correlated with this axis but do not contribute much to its inertia are 3 and 9. Species 9 tends to be associated with lodgepole pine age class I. Species 3 seems to have a weak affinity with age classes II and III. Similarly, the second axis contributes almost 87% to the

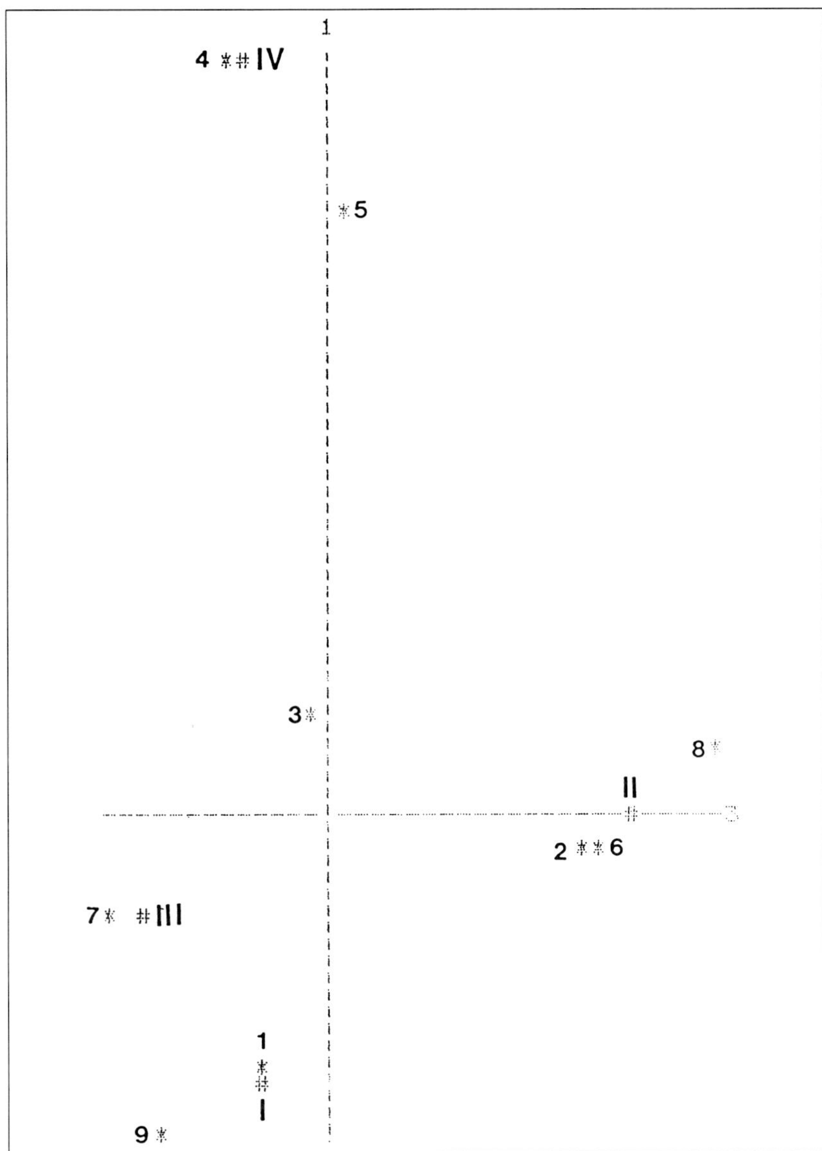

Figure 9.2 Scatterplot of points representing the fungi (rows) and age classes of lodgepole pine (columns) in Table 9.2 on the first and third correspondence analysis axes

inertia of age class III and 11% to age class I. The sum of the relative contributions of the axes to the inertia of a point (the squared correlations) is called the quality of that point's representation in the space and is unity for the total number of dimensions (i.e. the row sums of Table 9.5 are unity).

Table 9.4 Proportional contributions of the points to the inertias of successive axes

Fungal species	Axis		
	1	2	3
1	0.240	0.216	0.059
2	0.001	0.036	0.186
3	0.002	0.051	0.001
4	0.666	0.035	0.049
5	0.073	0.005	0.000
6	0.001	0.006	0.105
7	0.008	0.630	0.170
8	0.003	0.009	0.424
9	0.005	0.011	0.006
Sitka spruce age class			
I	0.231	0.301	0.062
II	0.000	0.040	0.708
III	0.014	0.621	0.190
IV	0.755	0.039	0.040

Table 9.5 Relative contributions of the axes to the inertias of the points (squared correlations)

Fungal species	Axis		
	1	2	3
1	0.609	0.353	0.038
2	0.011	0.326	0.662
3	0.069	0.927	0.004
4	0.950	0.032	0.018
5	0.956	0.043	0.000
6	0.018	0.124	0.858
7	0.018	0.888	0.093
8	0.029	0.053	0.919
9	0.389	0.506	0.105
Sitka spruce age class			
I	0.524	0.440	0.036
II	0.000	0.125	0.875
III	0.030	0.866	0.104
IV	0.956	0.032	0.013

Table 9.6 shows that fungal species 1, 3, 4, 5, 7 and 9 are well represented in two dimensions, having qualities greater than 0.895 (89.5%). However, species 2, 6 and 8 need the third dimension. Similarly, lodgepole pine age classes I, III, and IV are well represented in two dimensions, but age class II needs the third dimension.

The point which contributes most to the inertia in the third axis is lodgepole pine age class II (71%). Species 8, 2 and 6 contribute 42%, 19%

Table 9.6 Mass, quality, and inertia of the row and column points

Fungal species	Mass	Quality (2 axes)	Inertia
1	0.456	0.962	0.208
2	0.095	0.338	0.037
3	0.031	0.996	0.019
4	0.146	0.982	0.369
5	0.025	1.000	0.040
6	0.043	0.142	0.016
7	0.109	0.907	0.241
8	0.089	0.081	0.061
9	0.007	0.895	0.007
Sitka spruce age class			
I	0.406	0.964	0.232
II	0.253	0.125	0.108
III	0.175	0.896	0.244
IV	0.167	0.987	0.416

and 10% respectively (Table 9.4). Those points also have the largest squared correlations with the third axis (Table 9.5), the squared correlations of the remaining points being small. On the third axis (Figure 9.2), species, 2, 6 and 8 are associated with lodgepole pine age class II, and this group of points is separated from the rest on that axis.

This analysis may be compared with that of Dighton *et al.* (1986). They converted the counts to presence–absence data, and used logistic regression models to try to explain the relationships between tree age class and fungal species. They concluded that there was a succession with increasing age of lodgepole pine: *Laccaria   Paxillius, Lactarius, Cortinarius — Inocybe, Russula.* However, in these data *Laccaria* is absent from age class I and the analysis suggests a succession: *Lactarius, Paxillius — Inocybe, Cortinarius, Galerina, Russula — Russula, Laccaria — Clitocybe. Galerina* (species 8) is very closely associated with the moss layer developed under closed canopy stands. *Clitocybe* (species 4) is known to be associated with older stands. *Nolanea* is found under older stands but might be difficult to associate with one age class, its presence seems to be related to the accumulation of tree litter (Dr. J. Dighton, personal communication). The classes of such a succession cannot be delimited completely because of the tendency of some species to occur occasionally in another stage, e.g. *Russula* (species 3) occurs with reduced frequency in age class IV. Such a tendency can be represented on ordination charts such as those in Figures 9.1 and 9.2, which is one of the advantages of the method. Other advantages are that counts do not have to be transformed and the results, especially the visual displays, are easy to understand.

CHAPTER 10

COMPARISON OF ORDINATION METHODS

In order to discuss the different ordination methods, we need to introduce one which has not been discussed in detail, non-metric multidimensional scaling. This is one of a range of scaling methods, and it provides one way of avoiding the linearity constraints of principal component analysis without facing the choice of polynomial models or other non-linear functions. The method has not been discussed in detail here because although the principle is basically simple, the computer application is fairly complex. Interested readers are referred to Green and Rao (1972), Mather (1976), Springall (1978), Schiffman *et al.* (1982), Davies and Coxon (1982), and Coxon (1982), for more detailed explanations. The basic idea is that given a matrix of interobject dissimilarities (using one of a range of possible measures) it is required to represent them by an ordination (or mapping) into a space of reduced dimensionality. The basic assumption is that for a good ordination there should be a rank-order relationship between the original interobject dissimilarities and the distances in the ordination space, i.e. the relationship should be monotonic increasing. The exact functional form of this relationship need not be specified, which distinguishes the method from metric multidimensional scaling (Torgerson 1958) and principal co-ordinate analysis. In multidimensional scaling, solutions for different dimensionalities need to be computed separately as the solution for a space of given dimension is not necessarily a projection of solutions in higher dimensions.

Non-metric multidimensional scaling is related to other scaling methods such as principal co-ordinate analysis and correspondence analysis (Greenacre and Underhill 1982). A useful overview of these methods is given by Nishisato (1980). Such methods enable metric information to be extracted from strictly nominally-scaled attributes. The resulting metric scales often prove to be open to substantive interpretation (Gittins 1979 p460).

Greig-Smith (1983) discussed some applications of multidimensional scaling to vegetation data. Orloci (1978) suggested that the method has definite potential in revealing non-linear trends, although he noted the assumption that all points in the sample fall into a single cluster with even density. If there are two or more clusters, information will be lost in the course of the analysis. Gauch *et al.* (1981) compared four different non-metric multidimensional scaling programs in the analysis of real and

149

simulated plant community data. The value of non-metric multidimensional scaling in numerical taxonomy was shown by Rohlf (1972).

Multidimensional scaling has two major advantages in that it is non-linear and non-parametric. It has the special advantage that it seems to be better than principal component analysis in giving balance between the large intercluster distances and the small distances between cluster members (Rohlf 1970). However, there are three potential sources of weakness: (i) There must be a rank-order relationship between dissimilarity coefficients and distances, no permissible departure from monotonicity is known; (ii) A local optimum may be obtained in which there is an excellent fit between dissimilarities and distances, but the global fit may be poor; (iii) For good results, the true dimensionality must be known (Pimentel 1979). Problem (ii) is usually overcome by trying several different initial configurations (Chatfield and Collins 1980).

Rohlf (1972) made an empirical comparison of principal component analysis, principal co-ordinate analysis, and non-metric multidimensional scaling. Nine sets of zoological taxonomic data differing considerably in phenetic structure were used, ranging from random samples of individuals within a supposedly homogeneous population to species selected to represent variation in the order Lepidoptera. The criterion for comparison was the stress between the configuration of points in the ordination and in the original distance matrix. The equation for stress was

$$S = \left[\frac{\Sigma(D_{ij} - d_{ij})^2}{\Sigma(D_{ij} - M)^2} \right]^{1/2}$$

where D_{ij} is the distance between the *i*th and *j*th objects (individuals) in the reduced space, M is the mean of those distances, and d_{ij} is the value of D_{ij} expected assuming that there is a monotonic relationship between the D_{ij}'s and the original distances. Rohlf found that non-metric multidimensional scaling gave better results (lower stress) than did the other two methods, and he recommended its use unless N is large. Principal co-ordinate analysis gave a better fit than did principal component analysis when there were missing values, and he recommended that principal component analysis should be used only if there are no missing values and there are many more objects than attributes. He also noted that principal component analysis implies the use of Pythagorean distance only, whereas the other two methods allow the user to choose any distance measure including non-Euclidean ones.

In comparing non-metric multidimensional scaling, principal component analysis, and reciprocal averaging for the ordination of simulated coenoclines (community gradients) and coenoplanes (community patterns) using a stress criterion similar to that used by Rohlf (1972), Fasham (1977) concluded that non-metric multidimensional scaling produced better ordinations than did principal component analysis, and in most cases produced better results than did reciprocal averaging, provided that the non-metric multidimensional scaling was calculated using a space of the same dimensions as the simulated data. Non-metric multidimensional scaling was found to be less susceptible to distortion of the ordination by single gradients of high beta diversity and two-gradient situations when each gradient is of a different beta diversity (the degree of floristic difference among samples). Similarity measures evaluated for use with non-metric multidimensional scaling were: Pearson product-moment correlation coefficient, cosine θ coefficient, percentage similarity (Gauch and Whittaker 1972b), Kendall's rank correlation coefficient, Jaccard's similarity coefficient, and Manhattan distance. The cosine θ coefficient was found to give generally good results (cf. Q-mode factor analysis), the Manhattan metric produced inferior results.

Gauch *et al.* (1981) compared four different non-metric multidimensional scaling programs, reciprocal averaging, and detrended correspondence analysis in the analysis of real and simulated plant community data. They found that non-metric multidimensional scaling gave better results than did reciprocal averaging for data with three or four dimensions, but with only one dimension reciprocal averaging was better and for two dimensions there was little difference. Detrended correspondence analysis was considered to be superior to both the other methods. Hill and Gauch (1980) noted that when there is large habitat diversity, many of the larger distances will approach an upper limit for two samples with no species in common. Distances near this upper limit are all roughly equal with the result that while the local configuration of the ordination may be correct, the global configuration can be seriously distorted.

Brown *et al.* (1984) compared the use of detrended correspondence analysis and principal co-ordinate analysis on three sets of Tasmanian vegetation data having known gradients and one set where the vegetation was expected to respond to diverse environmental variables. In all cases, the results obtained by detrended correspondence analysis were considered to be superior to, or at least as good as, those of principal co-ordinate analysis.

Oksanen (1983) commented that it is difficult to compare ordination methods using real data which have partially unknown structure, reality is more complex than the simulations assume, and the species response model used in some simulations has been questioned. Also, the number of rare species is important. He used principal component analysis of covariance matrices, reciprocal averaging, detrended correspondence analysis, and both linear and non-metric multidimensional scaling to analyse relatively homogeneous data sets representing boreal heath-like vegetation in Finland. All the methods performed reasonably well, as was expected for rather short coenoclines. However, the ordinations differed in detail, revealing different aspects of the data sets. Reciprocal averaging and detrended correspondence analysis gave ordinations in which the species had nearly equal weight, apparently resulting from the double standardization. Multidimensional scaling and principal component analysis gave results which were determined chiefly by a few dominant species, and when the data were standardized so that the mean of positive occurrences was the same for each species while quantitative differences within species were retained, then the results from these methods were very similar to those obtained using reciprocal averaging and detrended correspondence analysis. Reciprocal averaging and principal component analysis were generally very good and reliable with these data provided that the impact of rare species and outlier relevés was removed in the former. Detrended correspondence analysis was found to be slightly less reliable than reciprocal averaging, presumably because the former was developed for use with very heterogeneous data. Multidimensional scaling was sensitive to uneven sampling patterns and was the least reliable method used.

Gauch *et al.* (1977) compared the use of principal component analysis (using several methods of centring, standardization and transformation of the data), polar ordination (Wisconsin comparative and Bray-Curtis), and reciprocal averaging, with field data and with simulated coenoclines and coenoplanes. The results were evaluated by comparing recovered structure (sample and species locations) with the original structure. They found that non-standardized principal component analysis was the most vulnerable to effects of beta diversity, giving distorted ordinations of sample sets with three or more half-changes (a measure of beta diversity). Polar ordination and reciprocal averaging gave good ordinations to five or more half-changes, and principal component analysis with standardized data was intermediate. Sample errors affected all these techniques more at low than at high beta diversity, but principal component analysis was the most vulnerable. Reciprocal averaging and polar ordination were little affected by the involution of axes extremes that affects non-standardized principal component analysis. Despite the 'arch' effect, reciprocal averaging was considered to be superior to principal component analysis at high beta diversities. None of these techniques solved all the problems that result from curvilinear relationships of plant community data. Outliers did not affect polar ordination unless they were used as end-points, but they can cause marked distortion in reciprocal averaging and principal component analysis. Clusters of samples tended to distort principal component analysis in rather unpredictable ways, but they have smaller effects on reciprocal averaging and more on polar ordination.

Pakarinen and Ruuhijärvi (1978) compared principal component analysis and reciprocal averaging in the ordination of Finnish peatland vegetation. They concluded that the relative merits of the technique depend upon the criteria used in their evaluation. Principal component analysis was considered to be better for presenting more than two axes of variation. Other authors have found the higher axes in reciprocal averaging to be variable and difficult to interpret (Bouxin 1976; Gauch *et al.* 1977). For species ordination reciprocal averaging gave the more interpretable results. Beals (1984) discussed the ordination of plant and animal species in terms of three types of ecological space: (i) samples as points in compositional or sociological space, (ii) samples as points in environmental space, and (iii) species (or other taxa) as points in niche space (resource or habitat). He noted that mathematical efficiency is not the same as ecological informativeness, and was especially critical of ordination methods involving centring of the data, because the centroid cannot exist in nature and the rotation is therefore a distortion of the environmental space and the sociological domain.

Beals (1984) noted that many commonly-used ordination methods do not maximize overall ecological information, but rather obtain specific information about the data set. Although reciprocal averaging has a perfect correspondence of a species ordination and a sample ordination and can handle longer gradients than most ordination techniques, it has substantial disadvantages, notably that although the first axis is good, successive axes are difficult to interpret. Detrended correspondence analysis removes the unwanted 'arch effect' in reciprocal averaging by drastic measures which Beals (p 16) likened to using a hammer to pound out an unwanted bulge. The question of why reciprocal averaging causes severe curvature is avoided rather than answered. Also, the reciprocity of species and sample ordination is lost. Non-metric scaling was considered to have the disadvantages that important information may be lost (through considering rankings rather than actual distances), that there may be more than one equally good configuration based on rankings, and that discontinuities may be masked.

Beals (1984) went on to discuss Bray-Curtis ordination in detail. The essence of the original method is: (i) calculate a distance matrix, (ii) select two reference points (either real or synthetic samples) for determining the direction of each axis, (iii) project all samples onto each axis by their relationships to the reference points. He discussed criticisms of the method and possible improvements, and the question of presence–absence versus quantitative data. In general, he considered Bray-Curtis ordination to be better than reciprocal averaging when more than one axis is needed, better than detrended correspondence analysis, and much better than principal component analysis. Non-metric scaling shows promise, but involves a much greater computational load.

Kenkel and Orloci (1986) noted that comparisons of ordination techniques often tend to confound three factors: (i) the methodological algorithm, (ii) the resemblance measure used, and (iii) the standardization used. They studied (a) the behaviour of principal component analysis, principal co-ordinate analysis, reciprocal averaging, and non-metric multidimensional scaling when the three factors mentioned above were not confounded; (b) the effect of data standardization on the results of metric and non-metric scaling; and (c) the utility and possible advantages of non-metric scaling in examining vegetation data.

The results obtained with simulated coenoplane data showed that prior data standardization had important effects on the results of all the methods used, although the effect depended on the method. Non-metric multidimensional scaling based on Pythagorean distance following stand norm standardization was found to be the best method for recovering simulated coenoplane data. Reciprocal averaging was considered to be the better of the other methods used, but in some cases it led to a distortion of results. The authors concluded that none of the currently-available ordination methods was appropriate in all circumstances.

The comparisons discussed above are all in the field of vegetation analysis, in which the aim is usually to arrange stands of species along a (usually hypothetical) environmental gradient. When a vegetation sample characterized by specific non-linear trends is ordinated on the basis of a linear or unfitting non-linear model, a certain amount of distortion can be expected. Types of distortion were discussed by Orloci (1978) who noted that among the different methods, principal component analysis stands out as the single most efficient technique for summarizing continuous linear data structures, although it is not very helpful if the data structure is non-linear.

Wilson (1981) presented statistical estimates of accuracy (the tendency of ordination results to mirror correctly the true community structure) and consistency (the ability of the ordination to produce concordant results) with tests for choosing among ordination techniques. The tests were used to evaluate ordinations of herbaceous vegetation data using detrended correspondence analysis and principal component analysis. For these data, detrended correspondence analysis was more accurate but principal component analysis was more consistent. For both methods the results suggested a sample size dependence. Wilson pointed out that an investigator should always seek corroborative evidence for ordination results.

Turning now from the preoccupations and problems of vegetation analysts, an interesting study was described by Gould (1969). In Bermuda, the snail *Poecilozonites bermudensis* underwent rapid evolutionary changes during the Pleistocene, giving rise to four distinct paedomorphic lines which could be regarded as subspecies. A factor analysis followed by varimax rotation produced four interpretative axes, the first two of which divided the snails into paedomorphs and non-paedomorphs and separated one of the paedormorphic species from the remainder. The third and fourth axes allowed much finer resolution of the situation to be obtained and stressed the subdivision of fossil and recent paedomorphs. When Blackith and Reyment (1971 p 206) tried a principal co-ordinate analysis on the same material, the sub-division of fossil and recent paedomorphs was not obtained.

Jöreskog *et al.* (1976) used Imbrie's Q-mode factor analysis, principal co-ordinate analysis, and correspondence analysis on a set of artificial data in which ten attributes for three objects were mixed in various proportions to create another seven objects. Only Q-mode factor analysis with oblique rotation gave an immediate representation of what was entered. Correspondence analysis gave a correct answer, but in a form which required considerable adjustment to obtain the real situation. Principal co-ordinate analysis using Gower's range-standardized Manhattan metric gave a distorted answer but the same method using Pythagorean distance would be expected to give a correct result.

Different methods illuminate different facets of the data. Chardy *et al.* (1976) recommended carrying out several types of ordination on the same data so as to identify structures of complementary interest. Dale (1975) commented that formal methods should be accepted or rejected simply on the grounds of whether they contribute to the problem solution or not.

CHAPTER 11

INTRODUCTION TO CLUSTER ANALYSIS

'Anything exact in clustering is rare and valuable'
(Hartigan, 1977)

In cluster analysis, little or nothing is known about the category structure, all that is available is a collection of observations whose category memberships are unknown. The objective, therefore, is to discover a category structure which fits the observations. The partitions of the category structure should have various desirable properties connected with 'naturalness', ease of manipulation, and the retrieval of information (Jardine and Sibson 1971; Anderberg 1973; Sneath and Sokal 1973; Clifford and Stephenson 1975). In seeking structure in the data, three possibilities should be borne in mind: (a) the data may contain no clusters, i.e. the points are uniformly distributed in the measurement space and lack cohesion; (b) the data may contain only one cluster, i.e. there is a high mutual association among all points; (c) a range of other possible configurations may fall between those two extremes. Again, in searching for structure in the data, it should be borne in mind that any given set of data may admit of several different but meaningful partitions, each of which may pertain to a different aspect of the data. Furthermore, cluster analysis can be a method for generating hypotheses. There is, as yet, no satisfactory definition of a cluster, and a partition obtained from a cluster analysis procedure has no inherent validity, its worth and its underlying explanatory structure are to be justified by its consistency with known facts.

Cluster analysis methods involve a mixture of imposing a structure on the data and revealing that structure which actually exists in the data. To a considerable extent, a set of clusters reflects the degree to which the data set conforms to the structural forms embedded in the clustering algorithm (Anderberg 1973). Many methods of cluster analysis break down for particular types of cluster that are, nevertheless, obvious to the eye. For example, elongated clusters are not well suited to Wishart's modal analysis, nor to methods that minimize $\mathbf{W}^{-1}$ (see Chapter 14); clusters with different dispersion matrices and outliers are not suited to methods which assume, explicitly or implicitly, homogeneity of within-group dispersions. Many methods tend to give clusters of approximately equal size, and failure to detect and eliminate an outlier can distort the procedure (Marriott 1974).

Jardine and Sibson (1971) pointed out that it had gradually been realized, in the previous few years, that despite superficial differences some of the variety of clustering algorithms which have been proposed implement the same method, and that different methods differ very widely in their properties and results. They also stated that the development of a general theory of cluster analysis had been hindered by two widespread confusions. The first of these confusions is between algorithms and the methods which they implement. Thus, Lance and Williams (1967a) suggested as a general theory of hierarchical clustering what is, in fact, a generalized agglomerative algorithm, for the distinction is correctly applied to algorithms rather than methods. The second confusion concerns the role of models in data simplification. The term 'model' may be used in two quite different ways. One use covers the mathematical framework within which it is possible to analyse the properties of the methods of data simplification. The other use covers descriptions of algorithms in terms of their application to some interpretations of the data. The latter may be called 'analogue models'.

Two kinds of analogue model have been widely used in cluster analysis. First, there are models which treat the objects as points or unit masses in Euclidean space (e.g. Gower 1967a; Wishart 1969a). Secondly, there are models which treat the objects as vertices of a graph, and values of the dissimilarity coefficient less than or equal to some threshold values as edges (e.g. Estabrook 1966; Jardine and Sibson 1968). Geometrical models can be applied only if the data are metric. Even when the data are naturally metric, a plausible geometrical interpretation for an algorithm does not necessarily provide any justification for the method which it implements. Thus, the various average-link and centroid algorithms which have simple geometrical interpretations suffer from very serious defects (Chapter 13). The graph-theoretic models are more generally applicable, since any dissimilarity coefficient and any stratified clustering can be characterized by a sequence of graphs (Chapter 15).

Most clustering methods are sequential, i.e. involve the application of a recursive sequence of operations to the set of objects that is considered to be the disjoint partition (agglomerative methods) or the conjoint partition (divisive methods). Simultaneous clustering methods have not found much favour.

An important division is between hierarchical and non-hierarchical methods. In hierarchical clustering, a partition at a given level of similarity or dissimilarity may be subdivided into smaller groups, or form part of a larger group, at other levels. The end-point of this process is a dendrogram, or tree diagram, in which numerical levels are associated with the branch points. The clusters specified at a particular level are generally non-overlapping, i.e. distinct clusters do not meet and every object belongs to only one cluster, possibly consisting of that object alone. Overlapping methods, in which some objects may belong to two or more groups at a given level, are generally non-hierarchical.

A hierarchical method optimizes a route between the objects of a sample, via intermediate groupings, to a single group consisting of the entire sample, by a series of fusions (agglomerative method) or in the reverse direction by a series of fissions (divisive method). In practice, a fusion or fission is generally pairwise, i.e. it involves two groups, two objects, or a group and an object. In non-hierarchical methods, the structure of the individual groups is optimized and the groups are made as homogeneous as possible.

Lance and Williams (1967a) defined a compatible strategy as one in which the metric among coarser clusters is the same as that among finer clusters or even among the original objects. Clustering methods which do not have this property are called incompatible and are considered to be undesirable.

Most clustering procedures are what Williams (1971) called 'intrinsic', i.e. the attributes measured on the objects decide the groupings. In 'extrinsic' clustering an external attribute is described in advance, and the resulting groups, although based on the internal attributes, are required to reflect discontinuities in the external attribute as closely as possible, the intrinsic pattern being overridden if necessary (e.g. Macnaughton-Smith 1965).

Hierarchical and non-hierarchical methods

A hierarchical agglomerative method must optimize some function between the two groups to be fused next, or between these two groups jointly and their fusion product, and similarly with fissions. The groups through which the process passes are not necessarily optimal in themselves, the best route may be obtained at the expense of some slight reduction in homogeneity of the individual groups.

Williams (1971) divided non-hierarchical methods into two types. In the first type, serial optimization, a group is defined and removed from the population. Within the remainder, a second group is defined and removed, and so on. The process may end with a final reallocation check. Strategies of this type are generally open to criticism on numerical or computational grounds. In the second type of non-hierarchical strategy, the groups are optimized simultaneously, and the model is always strictly Euclidean. However, there is reason to believe that the basic model is lacking in power, and the type of classification produced is not that commonly required by users. In non-hierarchical strategies no route is defined between groups and their constituent individuals, or between groups and the complete sample.

Marriott (1974) stated that if there is no special reason for imposing the nested structure of the dendrogram, the strictly hierarchical methods have serious disadvantages. For example, to decide whether a division into two or three groups gives a better representation of the data, it is necessary to compare the best division into two with the best division into three, and a hierarchical method will not necessarily give both. On the other hand, Blackith and Reyment (1971, p277) stated that it seems likely that although hierarchical techniques are almost always undesirable in theory, the consequences of using hierarchical techniques when the structure of the experiment renders such a practice dubious, seem not to be very serious. However, Hawkins *et al.* (1982 p307) gave an example which illustrates the danger of using a hierarchical clustering method on data for which that approach is not the logical choice.

Agglomerative and divisive methods

Beginning with a set of N objects, agglomerative methods group them into successive partitions, each having fewer than N groups, eventually arriving at a single group containing all N objects. Most of the widely-used clustering methods are agglomerative. By contrast, divisive methods commence with all N objects in one group and subdivide it into successive partitions until some stopping rule operates. In general, agglomerative methods are polythetic, while divisive methods tend to be monothetic.

Agglomerative methods are said to be prone to a small amount of misclassification because the process begins at the interobject level, where the information available is only that due to the pair of objects being considered. Clustering thus depends on the accuracy of the interobject measures where the possibilities of error are greatest. However, Sneath and Sokal (1973 p209) considered that many sequential agglomerative techniques reliably estimate interobject similarities within a cluster, but become increasingly unreliable as larger clusters are considered.

With divisive methods, there is a greater danger of inappropriate allocation of some objects that cannot be corrected later unless some special terminal reallocation procedure is used (Williams and Dale 1965). Monothetic divisive classifications have the advantage of simplicity and clarity of the group definitions, and new individuals can be allocated quickly and unambiguously to an existing group. However, it is a characteristic of such methods that two objects, closely similar in other respects, may be separated on the basis of a single attribute in which they differ. This troublesome form of misclassification means that such methods tend to produce an unduly large number of fragmentary groups (Williams 1971 p310).

Monothetic methods waste information, and there are severe practical problems in implementing divisive polythetic techniques (Pielou 1969). It is not surprising, therefore, that agglomerative polythetic methods have proved to be the most widely used.

Classification versus dissection

The objective of dividing a collection of objects into a number of groups immediately poses two problems:

(1)   Is there a 'natural' subdivision of the objects into groups? Such a subdivision may take one of two forms. Firstly, the groups may be distinct and defined by discontinuities in multivariate space. Secondly, and, perhaps, more usually, the tails of the frequency distributions of the groups may overlap and the groupings are likely to be based on multimodality (Marriott 1974). A given collection of objects may have groupings of both types. In such cases as these, the problem is one of *classification.*

(2)   A totally different problem occurs if the data are unimodal, then the question is what is the 'best' way of dividing the objects into groups. This was called *dissection* by Kendall and Stuart (1968 p314). It is important not to confuse this process with classification. In dissection there is no implication that the resulting groups represent in any sense a 'natural' division of the data, they are merely a matter of convenience and the only real criterion is their utility.

Classification involves the recognition of similarities between, and the grouping of, objects and organisms. A classification may have more than one purpose, but the paramount purpose is to describe the relationships of objects to each other, and to simplify the relationships so that general statements can be made about classes of objects. An important distinction is between monothetic and polythetic classifications. Monothetic classifications are those in which the classes established differ by at least one property which is uniform among the members of each class. In polythetic classifications, groups of individuals or objects share a large proportion of their properties, but do not necessarily agree in any one property. A corollary of polythetic classification is the requirement that many properties be used to classify objects. However, once a classification has been established, few characters are generally necessary to allocate objects to the proper groups. Hence, classification of a data set results in a reduction of the amount of information that is necessary to describe the data, but, if the classification is efficient, there is little or no reduction in the amount of information contained in the data. Classifications based on many properties will be general, they are unlikely to be optimal for any single purpose, but might be useful for a great variety of purposes. By contrast, a classification based on a few selected properties might be optimal with respect to those properties, but would be unlikely to be of general use (Sokal 1974).

If dissection is to be carried out, the basis of the dissection must be clearly defined. For example, an ecologist may regard the vegetation as essentially continuously changing, but changing more rapidly in some regions than others. He/she will therefore wish to treat these zones of maximum gradient as if they were discontinuities, and to sharpen them by an appropriate technique (Williams 1971). Some cluster methods have properties which make them useful for different types of dissection, for example the various minimum-variance methods and methods of the association analysis type. The flexible clustering strategy of Lance and Williams (1967a) may also be useful for this purpose.

All collections of objects can be dissected, not all can be classified. Classification may be a technique for generating hypotheses, but dissection is not, as the data are forced into a strait-jacket which restricts the domain of possible hypotheses and suggests that some will be generated by the process of dissection, rather than by the data (Cormack 1971).

CHAPTER 12

SIMILARITY AND DISSIMILARITY MEASURES

Here we shall consider similarity or dissimilarity between objects; similarity or dissimilarity between an object and a group of objects, or between groups of objects, will be considered under the various clustering methods. The difference between similarity and dissimilarity measures is that similarity measures increase with increasing likeness of objects, whereas dissimilarity measures decrease with increasing likeness. Thus, dissimilarity is the complement of similarity. In ordination and clustering methods, dissimilarities do not always correspond to distances, so the distinction is made.

A wide variety of measures has been proposed (Cormack 1971; Jardine and Sibson 1971; Sneath and Sokal 1973; Clifford and Stephenson 1975; Gower 1985). Some reflect the need to accommodate particular forms of data, as, for example, those restricted to binary data. Others allow for unevenness in the frequencies of attributes and minimize the influence of large or small values. Yet others are based on prior hypotheses concerning the statistical distributions of the properties measured. For a wide range of data, most of the more generally-used measures are monotonic with respect to one another, and many can be interconverted. However, they are not all interchangeable. Many measures have become neglected because they are mere variants of others or because they have undesirable properties. The correct choice of a dissimilarity or similarity measure in a given situation is extremely important. The choice is limited by the nature of the original observations, and, in general, the choice of algorithms decreases in the order: continuous, ranked, binary attributes and is most limited for disordered multistate attributes (Sneath and Sokal 1973). Gower (1970) discussed some of the problems of weighting characters and of eliminating correlation effects.

Dissimilarity can be defined in many ways, but the different dissimilarity measures do not all obey the rules of Euclidean geometry. A p-dimensional Euclidean space is a space consisting of all sets (objects, points) of p numbers $(x_1, x_2, \ldots . x_p)$ where the distance $d\ (x, y)$ between $x = (x_1 \ldots \ldots x_p)$ and

163

$y = (y_1 \ldots\ldots y_p)$ is given by:

$$d(x, y) = \left[\sum_{i=1}^{p} (x_i - y_i)^2 \right]^{1/2}$$

Williams and Dale (1965) pointed out that models in Euclidean space have at least three advantages: (i) many simple, robust and powerful methods are available for dealing with Euclidean systems that are not available for non-Euclidean systems; (ii) they satisfy the requirement for hierarchical classification that each level in the dendrogram is associated with some measure which shall decrease as the hierarchy descends; (iii) it is easier to gain intuitive perception of Euclidean systems and to grasp their properties, and to predict those properties in extreme cases. Another useful property of Euclidean space is that the distance between any two points is unaltered by orthogonal rotation of the co-ordinate axes. There are three obvious ways in which Euclidean properties may be lost: (i) p may not be constant, so that the dimensions vary locally; (ii) the p-tuples $(x_1 \ldots\ldots x_p)$ may be constrained in some way, e.g. to the surface of a sphere; (iii) the distance function may fail, it may hold within sets but may fail between some or all of them; the space is thus locally Euclidean and hence, if varying continuously, Riemannian. Here, no difficulty arises unless interset functions are required. Williams and Dale discussed three methods in which this type of problem arises, involving Mahalanobis' D^2, the method of Macnaughton-Smith *et al.* (1964), and association analysis.

Dissimilarity coefficients are required to satisfy the following conditions (Jardine and Sibson 1971): if d_{ij} denotes the dissimilarity between objects i and j, then

(i) $\qquad\qquad\qquad\qquad d_{ij} \geqslant 0$ for all i, j

(ii) $\qquad\qquad\qquad\qquad d_{ii} = 0$ for all i

(iii) $\qquad\qquad\qquad\qquad d_{ij} = d_{ji}$ for all i, j

Condition (i) implies that non-identical objects may or may not be distinguishable by the coefficient, condition (ii) states that identical objects are indistinguishable. Condition (iii) is the important symmetry relationship. An additional condition which may, or may not, be satisfied is

(iv) $\qquad\qquad\qquad\qquad d_{ij} + d_{jk} \geqslant d_{ik}$

This is the triangular inequality. A dissimilarity measure which satisfies all four conditions with the restriction that $d_{ij} > 0$ is called a metric or distance. Euclidean distances satisfy all four conditions, but not all metrics have Euclidean properties. If a dissimilarity coefficient has a value of zero for two objects which are known to have different attribute values, it is said to be pseudometric or semimetric. Condition (iv) may be replaced by

$$d_{ik} \leqslant \max(d_{ij}, d_{jk})$$

which is called the ultrametric inequality, and dissimilarity coefficients satisfying these conditions are said to be ultrametric. It should be noted that measures which are fully metric for complete data may become non-metric if there are missing data (e.g. see Clifford and Stephenson 1975 p81).

The most widely-used measures are discussed briefly below. For further details, and for other measures, see the books cited above, especially Sneath and Sokal (1973), and Williams and Dale (1965). It is convenient to group these measures under the headings: (a) Distance measures; (b) Association coefficients; (c) Probabilistic and information theory coefficients; but the boundaries between them are not sharp.

Distance measures

Minkowski metrics are a class of metric distance functions applicable to continuous attributes (interval or ratio scale data). The distance between any pair of points i, j over p attributes is given by

$$d_{ij} = \left[\sum_{k=1}^{p} |X_{ik} - X_{jk}|^r \right]^{1/r}$$

If $r = 1$ the measure is the Manhattan or city block metric, if $r = 2$ it is Pythagorean distance.

Pythagorean distance is the simplest and oldest of such measures. It is assumed that the p measurements used in calculating Pythagorean distance can be regarded as co-ordinates on a set of rectangular Cartesian axes in Euclidean space. Then

$$d_{ij} = \left[\sum_{k=1}^{p} (X_{ik} - X_{jk})^2 \right]^{1/2}$$

As the distance increases with the number of attributes, an average distance is often calculated as

$$d_{ij}(\text{av}) = d_{ij}/p \text{ or } \sqrt{(d_{ij}^2/p)}$$

This is particularly desirable if there are missing values, but missing values destroy the Euclidean properties.

Sometimes, in numerical taxonomy, d^2 is used as a distance measure because it is regarded as a desirable property of such measures that they should be additive over attributes, which d^2 is but d is not. Again d^2/p may be used as d^2 may become large with many attributes. However, squared Pythagorean distance is not a metric (Chatfield and Collins 1980).

Clearly, Pythagorean distance is unlikely to have much meaning if the attributes are in different units. Furthermore, it depends on the scale of the attributes, and will be dominated by attributes which have a greater range of values than others. It is therefore better used only when all the measurements are of the same type, or when they have been standardized in some way, e.g. by the variances.

Some authors (e.g. Blackith and Reyment 1971 p36) state that if two attributes are correlated, the Pythagorean distance between two objects calculated using those attributes will be over- or under-estimated by an amount which depends on the sign and magnitude of the correlation coefficient. Mather (1976 p314) gave a formula for calculating Pythagorean distance taking account of the correlations between attributes. Alternatively, it is suggested that the distances can be calculated using the components from a principal component analysis. However, it is difficult to reconcile this view with the fact that principal component analysis is a distance-preserving rotation, i.e. the interobject distances calculated using the component values are the same as those using the original data suitably centred or standardized.

Indeed, correcting for correlations may remove information of value. Gower (1972) suggested that information in correlated attributes could be useful in setting up groups, but once groups have been set up, the construction of discriminant functions requires the removal of correlations, hence the use of Mahalanobis' D^2.

Pythagorean distance can be calculated for binary data. With two attributes, two objects differing in only one of the attributes would be one unit distance apart; if they differed with respect to both attributes, they would be $\sqrt{2}$ units apart. Addition of further attributes produces a hypercube with sides of unit length. Since we are dealing with binary attributes, the calculation of the distance reduces to

$$d_{ij} = \sqrt{(b + c)}$$

where b and c are the numbers of attributes with different scores for objects i and j, i.e. the number of sides of the hypercube to be considered in calculating the diagonal. As with continuous data, comparisons involving different numbers of attributes should be divided by the number of possible comparisons.

Burr (1968) showed that Pythagorean distance can be calculated even if the attribute set is a mixture of continuous and disordered multistate attributes, although no comparable solution has been found for ordered multistate attributes (Williams 1976).

In some cases, Pythagorean distance is not a satisfactory dissimilarity measure. For example, if the measurements are all lengths the distance between two points will be largely a function of size. However, the direction of a point from the origin is an indication of shape, which might be of greater interest. In such a case the cosine θ coefficient (see Q mode factor analysis) would be more appropriate (Marriott 1974). Gower (1967b) noted that when the rows of the data matrix are ranked they have constant sum and constant sum of squares. With constant row sum, all sample points lie in a hyperplane, and with constant row sum of squares they lie on a hypersphere. In the latter case, the great circle distance between two points may be considered a more natural measure than Pythagorean distance. Again, this suggests the use of the cosine θ coefficient. A cosine is converted to a distance by (Boyce 1969):

$$d_{ij} = [2(1 - \cos \theta)]^{1/2}$$

The Manhattan metric is also quite widely used. If A (X_1, Y_1) and B (X_2, Y_2) are points on buildings in a town plan with streets at right angles, the distance from A to B is given by

$$d(A, B) = |X_2 - X_1| + |Y_2 - Y_1|$$

In general

$$d_{ij} = \sum_{k=1}^{p} |X_{ik} - X_{jk}|$$

We use the modulus, or absolute value, to avoid sign reversal if the distance is measured in the opposite direction. Manhattan metrics are not generally invariant under rotation of the attribute space, e.g. in ordination.

As with Pythagorean distance, some form of standardization is preferable, and three forms are in general use: the Gower metric, the Bray-Curtis measure, and the Canberra metric. In the Gower metric, standardization is by range (Gower 1967b, 1971a).

$$d_{ij}(\text{Gower}) = \sum_{k=1}^{p} \left[\frac{|X_{ik} - X_{jk}|}{\text{range } k} \right]$$

It is not suitable for highly-skewed data. If this metric is divided by p, it is constrained between 0 and 1. Its square root is a Euclidean distance, this holds when the standardizing constant is equal to, or greater than, the range (Gower 1972).

The Bray-Curtis measure, which should only be used if the data are everywhere non-negative is

$$\frac{\sum_{k=1}^{p} |X_{ik} - X_{jk}|}{\sum_{k=1}^{p} (X_{ik} + X_{jk})}$$

In this measure, which is constrained between 0 and 1, the denominator is a sum involving all properties of the two objects, and it tends to be greatly influenced by occasional outstanding values and attributes of large range. It is usual to pre-standardize the raw data before applying the measure (Williams and Lance 1977). It is thought that the Bray-Curtis measure is not a metric (Lance and Williams 1967b). Beals (1984) considered the measure to have many advantages in vegetation studies. Applied to binary data, this collapses into the ones-complement of the Czechanowski similarity coefficient (q.v.).

The Canberra metric, which should also be used only if the data are everywhere non-negative, is

$$d_{ij}(\text{Canb}) = \sum_{k=1}^{p} \left[\frac{|X_{ik} - X_{jk}|}{(X_{ik} + X_{jk})} \right]$$

This measure differs from the Bray-Curtis in being the sum of a series of fractions, so that an outstanding value can contribute to only one of the fractions and thus does not dominate the measure. It is a property solely of the objects or groups being compared, and is not affected by the range of the entire attributes. It is not sensitive to outlying values (Williams 1976) but is sensitive to proportional rather than absolute differences (Sneath and Sokal 1973). With vegetation data the result will be biased toward the constancy and fidelity of species in site groups (Clifford and Stephenson 1975 p90). However, the Canberra metric is sensitive to zero values in the data, since if one of a pair of values is zero the fraction becomes unity irrespective of the size of the other element. This problem may be avoided by replacing the zero value by a small positive number for that comparison and neglecting double-zero comparisons, averaging the measure over such comparisons as remain (Williams and Lance 1977). If this metric is divided by p, it is constrained between 0 and 1. Applied to binary data, this collapses into the ones-complement of the Jaccard similarity coefficient (q.v.).

The above distance measures are not equally suitable for all purposes and all types of data. In particular, they differ in their responses to outlying values and to zero values. Double-zero matches do not contribute to the Bray-Curtis measure, but they do contribute to the other three. Williams (1976) gave an example of the application of the above measures to an artificial data set of species numbers at three sites. The ratios of the distances obtained by variance-standardized Pythagorean distance were the same as those for the Gower metric, the ratios of the distances calculated by the Canberra metric were slightly different from those of the previous two measures, but the Bray-Curtis measure was dominated by one large value.

With binary attributes, a similarity based on $|X_1 - X_2|$ will, if standardization is by standard deviation, cause the presence of a rare attribute or the absence of a common one to make a disproportionately large contribution to the analysis (Williams 1971 p316).

Pythagorean distance, even with standardized data, will be influenced by strong outliers or extreme skewness. If the data were everywhere non-negative, with few zeros but with occasional extreme outliers which the classifier does not wish to dominate the analysis, the Canberra metric might be used. The selection of a distance measure requires careful thought. If the measurements are log lengths, then a logical dissimilarity measure is the sum of squares of differences taken from the mean difference (Marriott 1974).

The effect of including logically-correlated attributes, e.g. length, breadth, and weight, is to increase the effective weighting given to the basic attribute, in this case size, which they all represent. Some authors eliminate correlations by using Mahalanobis' generalized distance, but Gower (1969) warned that this could result in a loss of valuable information. Marriott (1974) noted that if a distance measure is to be used to determine a grouping, it is illogical to eliminate the correlations before the grouping is formed. There is thus little justification for using Mahalanobis' distance in cluster analysis except in the way described by Friedman and Rubin (1967) or Marriott (1971).

Faith *et al.* (1987) used simulated vegetation data to examine the relationships between various coefficients of compositional dissimilarity and ecological distances, particularly with regard to the relative robustness of coefficients to variations in the model of community variation in ecological space. The Bray-Curtis measure, a Manhattan distance standardized to equal totals, and another measure called the Kulczynski measure had not only a robust monotonic relationship with ecological distance, but also a robust linear (proportional) relationship until ecological distances became large. Among the less robust measures were chi-square, Pythagorean distance, simple Manhattan and Gower's range-standardized Manhattan metrics.

Similarity or association coefficients

There are many of these coefficients, which measure the agreement between pairs of objects over a range of binary, or sometimes multistate, attributes. Here, the more widely-used coefficients, which are constrained between 0 and 1, will be discussed. A coefficient so constrained can be converted into a dissimilarity by subtracting it from 1, and some can be converted into distance metrics. For a fuller discussion of the various coefficients and their properties, see Greig-Smith (1964), Cheetham and Hazel (1969), Blackith and Reyment (1971), Sokal and Sneath (1973), Orloci (1978), Wolda (1981), Hubalek (1982).

With binary data, the attribute-states of two objects i and j can be compared by a 2×2 frequency table:

<table>
<tr><td></td><td></td><td colspan="2">object j</td><td></td></tr>
<tr><td></td><td></td><td>1</td><td>0</td><td></td></tr>
<tr><td></td><td>1</td><td>a</td><td>b</td><td></td></tr>
<tr><td>object i</td><td></td><td></td><td></td><td>$p = a + b + c + d$</td></tr>
<tr><td></td><td>0</td><td>c</td><td>d</td><td></td></tr>
</table>

where a is the number of attributes coded 1 for both objects, and so on for b, c, d. The main problem in dealing with binary data is that of symmetry, i.e. whether or not to include joint absences. In many ecological problems it would be absurd to call two sites alike simply because they both lack a large number of possible species or to call two organisms alike because they both lack a number of possible characters. Faith (1983) discussed asymmetric coefficients.

Three coefficients commonly used for binary data are the simple matching coefficient, the Jaccard coefficient, and the Czechanowski coefficient. These will now be discussed in more detail. Missing values could be allowed for by changing the denominator as necessary, but any Euclidean properties of derived distances would be lost (Gower 1972).

The simple matching coefficient (SMC) is one of the oldest and simplest measures. It is defined as the total number of matches, both (1,1) and (0,0), divided by the total number of comparisons

$$\text{SMC} = \frac{a + d}{a + b + c + d}$$

and it takes values between 0 and 1. Its complement is

$$1 - \text{SMC} = \frac{b + c}{a + b + c + d} = \frac{d_{ij}^2}{p}$$

where d_{ij} is the Pythagorean distance based on the raw binary data states (Sneath and Sokal 1973; Clifford and Stephenson 1975).

The Jaccard coefficient is the simplest of the coefficients which omit joint absences, and it tends to emphasize difference. The number of (1,1) matches is divided by the total number of comparisons

$$\text{Jaccard} = \frac{a}{a + b + c}$$

and it takes values between 0 and 1. Its complement is

$$1 - \text{Jaccard} = \frac{b + c}{a + b + c}$$

which is not fully metric as it may fail to satisfy the triangular inequality (Clifford and Stephenson 1975; van der Maarel 1979). However, its square root is a Euclidean distance (Gower 1972). This coefficient is monotonic with the Czechanowski coefficient (Sneath and Sokal 1973).

In the Czechanowski coefficient (also attributed to Dice and to Sorensen), the number of (1,1) matches is divided by the arithmetic mean of the number of presences for the two objects. This becomes (Williams 1976)

$$\text{Czechanowski} = \frac{2a}{2a + b + c}$$

and it takes values between 0 and 1. Its complement is

$$1 - \text{Czechanowski} = \frac{b + c}{2a + b + c}$$

which is not fully metric (Williams and Dale 1965; Clifford and Stephenson 1975; van der Maarel 1979). The quantitative equivalent to this coefficient is the Bray-Curtis measure.

With some data, there may be little to choose between the Jaccard and the Czechanowski coefficients, they are jointly monotonic. If there are relatively few joint presences the Czechanowski coefficient has larger values, which might be considered an advantage. With relatively many joint presences, the Jaccard coefficient may be more attractive because it will give a wider spread of values near the upper end of the range (Clifford and Stephenson 1975).

Other coefficients are Simpson's which emphasizes similarity, rather than difference as in the Jaccard coefficient, and Otsuka's which is the binary equivalent of the cosine θ coefficient (see Hazel 1970). Lamont and Grant (1979) discussed 21 dissimilarity measures suitable for plant species presence–absence data. The best measure of difference in species composition between two sites is that which closely reflects their ecological distance along an environmental gradient. In a Monte Carlo study of data on moss floras of five regions of Newfoundland, Rice and Belland (1982) commented that the Jaccard and other binary similarity coefficients are sensitive to the sizes of both the total species pool available and the floras (or faunas) drawn from the total pool. They considered such coefficients to be unsuitable as direct reflections of the magnitudes of ecological or biogeographical affinities of sites.

Hubalek (1982) made detailed studies of 43 similarity coefficients. Some were found to be synonyms or direct correlates with indices described by earlier workers, others were mere transforms from one range of values to another. Several coefficients were incompatible with suggested admissibility conditions. In all, 23 coefficients were rejected and the remaining 20 were used in an empirical trial on interspecific association data among fungi of the genus *Chaetomium*, with the use of cluster analysis. The Jaccard and Czechanowski (Dice-Sorensen) coefficients performed well.

Multistate attributes can be treated as binary, as in Gower's general similarity coefficient which also overcomes the problems of mixed data types (Gower 1971a). In this coefficient, the similarity between the ith and jth objects over all the attributes is the sum of the individual similarities over the different types of attributes:

$$S_{ij} = \sum_{k=1}^{p} s_{ijk} w_{ijk} \bigg/ \sum_{k=1}^{p} w_{ijk}$$

where p is the total number of attributes. The normal use for w_{ijk} is to indicate whether a comparison is possible ($w_{ijk} = 1$) or not ($w_{ijk} = 0$). This coefficient is easier to understand if we consider the individual attribute types. For binary attributes we have the choice of allowing double-zero comparisons or not. If double-zero comparisons are not allowed, then for each occurrence of double-zero $w_{ijk} = 0$ and $s_{ijk} = 0$. Otherwise $w_{ijk} = 1$ and $s_{ijk} = 1$ for a match (1,1) and zero for a mismatch (1,0 or 0,1). This is clearly the same as the Jaccard coefficient. If double-zero comparisons are allowed, then for each occurrence of double-zero $w_{ijk} = 1$ and $s_{ijk} = 1$. This is clearly the same as the simple matching coefficient. Similarly, for multistate attributes $s_{ijk} = 1$ if objects i and j agree with respect to the kth attribute and equals zero if they differ.

For quantitative attributes

$$s_{ijk} = 1 - |X_{ik} - X_{jk}|/\text{range } k$$

which is the complement of the range-standardized Manhattan metric mentioned earlier. Over all p attributes, Gower's general similarity coefficient has values from 0 to 1. Gower (1971a) showed that the resulting similarity matrix is positive semi-definite if there are no missing values. This means that it can be used in principal co-ordinate analysis, and the distances given by $\sqrt{(1 - S_{ij})}$ can have a real Euclidean representation. This is also true for the individual types of similarity contained in the general definition.

Gower's general similarity coefficient has been used in taxonomic studies of organisms and in soil studies (Rayner 1965, 1966). Aitchison (1978) used it to analyse four numeric and three multistate attributes in a socio-economic study of the French landscape. Corresponding distances were calculated as

$$d_{ij} = \sqrt{(2(1 - S_{ij}))}$$

and were then analyzed using a minimum spanning tree clustering method and Ward's error sum of squares method. Radloff and Betters (1978) and Betters and Rubingh (1978) used the coefficient with UPGMA clustering to analyse data of mixed types for wildland classification and suitability analysis.

The Pearson product-moment correlation coefficient calculated between pairs of objects is sometimes used as a similarity coefficient in numerical taxonomy. In statistical theory, this coefficient estimates a parameter of a bivariate normal frequency distribution. When it is used as a similarity coefficient, it is calculated between the rows of a data matrix rather than the columns, i.e. across attributes. The heterogeneity of the elements of the row vectors does not match the statistical derivation of this correlation coefficient, and this application is inappropriate. Those who use the correlation coefficient in this way try to overcome the problem by standardizing the rows of the data matrix. This in itself is meaningless, because the basis of standardization is a normal distribution, and the row (object) vectors are not likely to have that. Gower (1967a) thought that it is difficult to justify the use of the Spearman formulae in calculating correlations between objects (see also Minkoff 1965; Everitt 1974 p 53).

The correlation coefficient implies a distance which is often considered to be Euclidean, but in fact constrains points to a semi-hypersphere and hence to a Riemannian spatial system (Beals 1984). However, if such a geometrical representation seems appropriate, the cosine θ coefficient can be used (see Q-mode factor analysis). That coefficient is proportional to the great circle distance between points, and is a measure of proportional similarity (Gower 1967b). Imbrie and Purdy (1962) described a hierarchical clustering method appropriate to this measure.

Probabilistic and information theory coefficients

Attempts have been made to develop similarity coefficients which could take into account the frequency distributions of the attributes, i.e. agreement among rare attributes is less probable than agreement among frequently-occurring attributes, and to weight accordingly. Goodall (1966a) developed a probabilistic similarity index capable of using binary, ordered multistate, and continuous attributes (Sneath and Sokal 1973; Clifford and Stephenson 1975). The use of the method requires a considerable computational load, and Sneath and Sokal expressed doubts about some aspects of this coefficient. The application of information theory concepts is also based on the frequencies or probabilities of the occurrence of the species among stands, i.e. on the likelihood of each species occurring in a random sample.

Information statistic measures are particularly applicable to binary and disordered multistate attributes, it has proved difficult to apply them satisfactorily to continuous attributes (Sneath and Sokal 1973; Clifford and Stephenson 1975; Williams 1976). Williams (1971) suggested that such measures are particularly applicable to highly-skewed binary data of plant species. Information measures have the advantage of handling missing data effectively (Clifford and Stephenson 1975).

The information content I of a sample, which can be calculated in various ways (Sneath and Sokal 1973; Clifford and Stephenson 1975; Williams 1976; Williams and Lance 1977), is a measure of its heterogeneity. The more heterogeneous the sample, the more information it contains. If two samples, A and B, are combined, the information content of the combination C cannot be less than the sum of the information contents of A and B separately; if it is greater, there is an information gain ΔI,

$$\Delta I = I_\mathrm{C} - (I_\mathrm{A} + I_\mathrm{B})$$

and ΔI can be used as a measure of dissimilarity between the samples. This method can be applied to both sets of objects and pairs of objects, although it is mainly used in clustering when either the gain in information ΔI can be minimized or the sets fused whose fusion has minimum I (Lambert and Williams 1966).

If it is assumed that the attributes are mutually independent, it is possible to estimate the probability that an observed ΔI arose by chance. $2\Delta I$ is then distributed approximately as chi-square with the degrees of freedom equal to the number of attributes (Clifford and Stephenson 1975). However, Field (1969) found that information analysis has undesirable group-forming properties when applied to a data matrix containing a large number of zero entries. With such data, Field showed that the degrees of freedom for chi-square should be one less than the number of attributes used, which means testing a different null hypothesis. However, Bottomley (1971) concluded that chi-square tests are not suitable for testing in conjunction with the information statistic.

CHAPTER 13

SEQUENTIAL HIERARCHICAL AGGLOMERATIVE NON-OVERLAPPING CLUSTERING METHODS

Group size dependence

Suppose we have a model which treats the objects as points or unit masses in Euclidean space. The initial dissimilarities define the space, but as groups fuse the updated dissimilarities may not define a space with the original properties. One type of strategy might simply draw boundaries around groups of points without changing the spatial relationships of the points. Such strategies were termed 'space-conserving' by Lance and Williams (1967a). Centroid clustering and flexible clustering with $\beta = 0$ are the only strategies which are completely space-conserving, but the group average strategy (UPGMA) is very nearly so.

In some strategies, as a group grows in size it becomes progressively more difficult to join. That is, the space between the group and other groups appears to expand. Such strategies were called 'space-dilating' by Lance and Williams (1967a) and they have the effect of sharpening discontinuities in the system so that the clustering is intensified. Clearly, this introduces a distortion, but if such strategies are used with an understanding of their properties they can be useful in dissection.

The space dilation effect involves the group size dependence of the group/group and object/group measures (see Williams 1971, 1976). In either case, the dependence may be asymptotic, i.e. once a group has attained a modest size further additions make little difference, or it may be indefinite, so that every addition makes the group substantially more remote and hence more difficult to join. Object/group measures may have different properties from the group/group measures of which they are a limiting case. In the flexible clustering strategy, both measures are asymptotic with β negative, the dilation being more intense the more negative the value of β. In the incremental sum of squares strategy the object/group measure is asymptotic while the group/group measure is indefinite. For at least one information statistic strategy, both measures are indefinite (Williams *et al.* 1971) and this is an intensely space-dilating strategy. Furthermore, if there are several large groups, each with one or two associated outlying points these points may not be able to join the groups with which they have the greatest affinity when their turn comes. This results in a collection of outlying points which are unlike all the groups and each other. The furthest neighbour strategy is also an intensely space-dilating one, but the degree of space dilation cannot be controlled by the user. If controlled space dilation is desired, flexible clustering is recommended. A value of $\beta = -0.25$ is found suitable for many purposes (Clifford and Stephenson 1975).

In space-contracting strategies, a group becomes easier to join as it grows, i.e. the space appears to shrink. In such strategies, clustering is very weak and chaining is common. This is particularly characteristic of single-linkage clustering or flexible clustering with β positive.

Clustering methods using dissimilarities

In sequential hierarchical agglomerative methods, the smallest dissimilarity between groups (a group may contain one object) is found and is taken as the current level. In pair-group methods only one pair of groups may be joined at any one time, but this constraint may be relaxed to give variable-group methods. When groups have been joined, new intergroup dissimilarities are calculated in some way, and the process is repeated until all the objects have been combined into a single group. The various methods differ in the way in which the new intergroup dissimilarities are calculated.

At any level, ties in the smallest dissimilarities may occur. If there are three groups, A, B, C, any of which may contain only one member, and if $d(A, B) = d(B, C)$, the program may have to choose which fusion to implement and will not necessarily indicate that a choice has been made. Usually, the first one encountered by the program is used arbitrarily. This may, or may not, affect subsequent clustering. Suppose that A and B fuse to form D, then the method for calculating $d(C, D)$ could result in C fusing with D at a later stage, possibly after other groups have been fused with D. This property of discontinuity has been used as a criticism of average-linkage procedures (Gordon and Birks 1972). Any errors introduced in this way will be especially important at the early stages of clustering.

Although these methods are discussed with reference to dissimilarities, many of them can also be used with similarities bearing in mind that dissimilarity values decrease for increasing likeness whereas similarity values increase. Greig-Smith (1983 p 204–205) gave a table of the properties of seven agglomerative strategies when used with five similarity and dissimilarity measures. Burr (1970) showed that when information gain is used with agglomerative strategies, the result may not be monotonic in certain, rather stringent, conditions.

The result of a hierarchical clustering algorithm can be both tabular and graphical. The tabular form lists the fusions at successive levels and the associated dissimilarity values. The same information can be represented diagrammatically in the form of a dendrogram, or tree diagram, in which dissimilarity (or similarity) levels are associated with branch points. A dendrogram which displays phenetic relationships among organisms is called a phenogram. A necessary and sufficient condition for an observed dissimilarity coefficient to be exactly represented by a dendrogram is that it should satisfy the ultrametric inequality, and the aim of a hierarchical clustering method is to find a tree such that the derived ultrametric dissimilarities are, in some sense, as close as possible to the observed dissimilarities (Chatfield and Collins 1980). The ultrametric inequality ensures that the pair-function implied by the dendrogram is monotonically increasing for dissimilarities or decreasing for similarities (Sneath and Sokal 1973). Lack of monotonicity is a serious defect.

The most commonly-used methods are: (1) Single-linkage (nearest neighbour); (2) Furthest neighbour; (3) Unweighted pair-group using arithmetic averages, called by Lance and Williams (1967a) the group-average method; (4) Weighted pair-group using arithmetic averages, (5) Unweighted pair-group centroid method, called by Lance and Williams (1967a) the centroid method; (6) Weighted pair-group centroid method, called by Lance and Williams (1967a) the median method.

Some confusion is possible over the use of the terms 'weighted' and 'unweighted'. Sokal and Michener (1958) wanted to give merging branches in a dendrogram equal weight regardless of the number of objects carried on each branch. Thus, if object k joins group (i, j) k and (i, j) are weighted equally. However, the individual objects obviously have unequal weights. Hence, in weighted clustering it is the individual objects that are weighted, according to the number of prior clustering steps in which they were involved. Conversely, in unweighted clustering the individual objects are unweighted, i.e. carry equal weights, but the clusters are effectively weighted by the number of objects in them (Sneath and Sokal 1973). This nomenclature is the reverse of that which one might normally use (Gower 1967a), but it has become established and is used here to avoid confusion.

In the single-linkage (nearest neighbour) method, new intergroup dissimilarities are not calculated. Instead, the original dissimilarities are retained, and clustering is based on the smallest dissimilarity between a point outside a group and a point inside the group. If this is implemented as a pair-group method and there are two or more equal smallest dissimilarities, an arbitrary choice is made as to which should be used. The resulting dendrogram is not affected by the choice (Chatfield and Collins 1980).

As a cluster expands, its outside members are nearer to the outside members of other clusters, and are thus more likely to link with them. Lance and Williams (1967a) described this method as 'space-contracting', and it is this property that is responsible for the effect known as 'chaining', i.e. clusters connected by intermediate objects are fused. Another form of chaining occurs when a small group forms at a low distance level, then expands by the addition of single points. This occurs where there is no real structure in the distribution of points in multidimensional space.

Williams *et al.* (1966) defined a coefficient of chaining, but Cormack (1971) commented that 'in the absence of a formal definition of chaining such a coefficient gives only spurious precision to any argument about the undesirability of this property'. Many authors consider this property to be sufficiently undesirable to render the method useless. However, that is an excessively harsh judgement because chaining simply indicates that there are no discontinuities in the data. Used thoughtfully, the single-linkage method is useful for exploratory work.

Certainly, the single-linkage method is conceptually and computationally very simple, and it has a large number of satisfactory mathematical properties. In particular, it does not suffer from dicontinuity; Jardine and Sibson (1971) criticised alternative hierarchical methods for their lack of continuity which they regarded as being a far more severe defect than chaining in many applications. The single-linkage method gives clustering of identical topology for any monotonic transformation of the distances. Jardine and Sibson (1971) proposed an axiomatic framework for clustering methods within which the single-linkage method is uniquely acceptable, and in that context its defects must be viewed as those of hierarchical clustering itself. Sibson (1973) stated that since the defects of the single-linkage method are well-enough understood and of such a nature as to cause it to be misleading only rather rarely, the method itself should be generally acceptable. The single linkage method is good for finding long and straggling clusters or those with unusual shapes.

An efficient way of computing single-linkage cluster analysis is via the minimum spanning tree (Gower and Ross 1969; Ross 1969b, c, d).

Furthest neighbour clustering is one of a number of algorithms for finding a subset of all possible complete-linkage clusters. Sorensen's (1948) method is another. Complete-linkage clusters are defined by the criterion that all links within a cluster must be of length equal to, or less than, the current level, so that two clusters join if the dissimilarity of the furthest pair of members, one in each cluster, is equal to, or less than, the current level.

The method generally leads to tight, spheroidal, discrete clusters that join others only with difficulty and at relatively large overall dissimilarity values. Lance and Williams (1967a) called this method 'space-dilating', Sneath and Sokal (1973) listed it as monotonic, but Jardine and Sibson (1971) pointed out that in this method the output dissimilarities are not continuous functions of the inputs. The effects of this discontinuity are not predictable in practice, and can lead to completely misleading results. The exact rules for handling ties are of some importance in deciding the form of the dendrogram in this and most of the other methods described here. Sibson (1972) criticised the complete-linkage method because it tries to pick out maximal complete subgraphs and at the same time generate disjoint clusters, which is impossible in general. The method is too reluctant to unite clusters.

To avoid the extremes of chaining on the one hand and, on the other, small, tight, compact clusters that leave out many of the less-easily affiliated objects, other clustering methods were developed. The average-linkage methods may be divided into the arithmetic average and the centroid methods.

Arithmetic average clustering computes the arithmetic average of the dissimilarity coefficients between an object candidate for admission and members of an extant cluster, or between the members of two clusters about to fuse. The arithmetic average may be unweighted or weighted, bearing in mind the definition given above that unweighted objects carry equal weights but the clusters are effectively weighted by the number of objects in them, whereas weighted objects are weighted according to the number of prior clustering steps in which they were involved.

In the unweighted pair-group method using arithmetic averages (UPGMA), also called the unweighted average-linkage method or the unweighted pair-group method and called by Lance and Williams (1967a) the group average method, all objects in a cluster are weighted equally. The distance between two groups a and b is defined as the arithmetic average of the distances between pairs of members (i,j) of both groups, i.e.

$$D_{ab} = \frac{1}{n_a n_b} \sum_{i=1}^{n_a} \sum_{j=1}^{n_b} d_{ij}$$

The method is regarded as space-conserving, having no marked tendencies to contraction or dilation (Lance and Williams 1967a), and the dendrogram is monotonic (Mather 1976). UPGMA is not invariant under an arbitrary, strictly monotone, transformation of the distances.

In the weighted pair-group method using arithmetic averages (WPGMA), also called the weighted average-linkage method, weighted pair-group method, and simple average method, the object most recently admitted to a group is weighted equally with all previous members, or if two groups are fused they are given equal weight irrespective of the number of objects in each. The distance between two groups a and b is the average of their interobject distances, i.e.

$$D_{ab} = \sum_{i=1}^{n_a}{}' \sum_{j=1}^{n_b}{}' w_i w_j d_{ij}$$

where $w_i = 1/2^c$ and c is the number of prior clustering steps of object i.

This method is also considered to be space-conserving (Sneath and Sokal 1973), however, in taking account of small groups the method tends to dilate intergroup distances (Mather 1976).

Details of these methods are given by Gower (1967a), Sneath and Sokal (1973) and Jardine and Sibson (1971). The general structure produced by UPGMA is similar to that produced by the complete-linkage method, but there are some fine distinctions. WPGMA shares the properties of UPGMA but distorts the overall relationships in favour of the most recent arrival in a cluster. This may tend to distort the space to make groups appear to be further apart than they appear when UPGMA is used (Mather 1976). Sneath and Sokal (1973) listed both of these methods as monotonic, but Jardine and Sibson (1971) pointed out that the output dissimilarity coefficients are not continuous functions of the inputs. If measures other than Euclidean distance are used, the concept of average dissimilarity needs some thought (see Lance and Williams 1967a).

Because the group average methods are weakly clustering (space-conserving) they are to be preferred in situations where the user does not wish to force a structure on the observations, or to check for misclassifications arising from the use of other methods.

Centroid clustering finds the centroid of the objects in a cluster, and measures the dissimilarity (usually Euclidean distance) of any candidate object or cluster from this point. Centroid clustering has a simple geometrical interpretation, but no similar geometrical interpretation can be found for arithmetic average clustering. The unweighted pair-group centroid method (UPGMC) weights all objects in a cluster equally. When two clusters join, the resulting centroid is nearer to the centroid of the larger of the parent clusters (Fig. 13.1). Lance and Williams (1967a) called this the centroid method and considered it to be space-conserving. However, the ultrametric inequality requirement is not met and the output dissimilarities are not monotonic, i.e. the dendrograms show reversals. The weighted pair-group centroid method (WPGMC) was called the median method by Lance and Williams (1967a) from its linear combinatorial formula first developed by Gower (1967a). This method weights the most recently-admitted object in a cluster equally to the previous members. When two clusters join the resulting centroid is midway between the centroids of the parent groups. This method shares many of the properties of UPGMC, including a lack of monotonicity, although there are some differences in the resulting dendrogram caused by the greater weighting given to the late joiners of clusters (Sneath and Sokal 1973).

Because of the lack of monotonicity, and consequent reversals in the dendrograms, the centroid methods have been largely avoided for some time, and the strategy may be regarded as obsolete.

Ward (1963) proposed a hierarchical agglomerative procedure in which the aim is to minimize the loss of information in passing from one level to

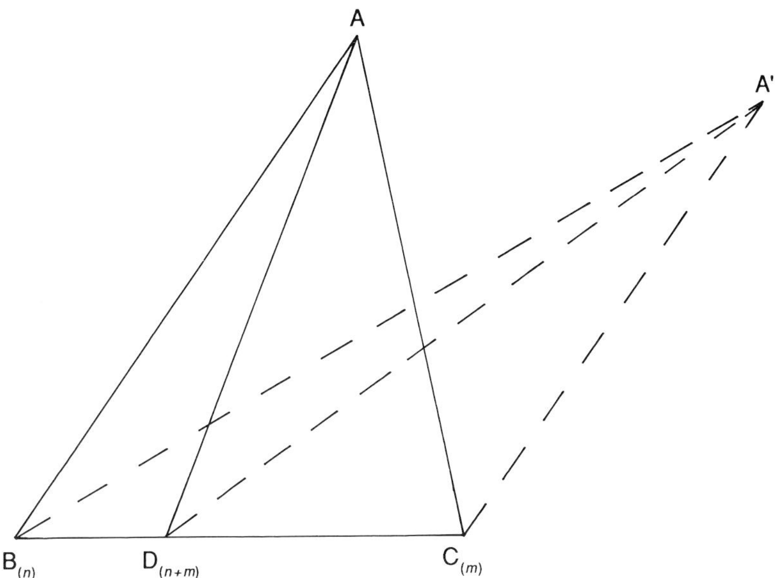

Figure 13.1 Centroid pair-group clustering. B and C are centroids of groups having n and m members respectively ($n > m$). BC is less than AB or AC. When B and C join, the centroid of the new group is D. In UPGMC, the ratio BD:DC is as $m:n$. In WPGMC BD = DC. In either case, the point now representing B is nearer to A than was the original point, and the point representing C is further away. The magnitude of this effect depends on the relative positions of the points; it is more pronounced if A is in the position A'

the next. If a given level contains k groups there are $k(k-1)$ possible fusions, and the one selected is that which minimizes the increase in the loss function. If two or more fusions yield the minimum value an arbitrary choice is made. In practice, the loss function is normally the sum of the squared deviations of all observations from their cluster means. With N objects, for a given attribute

$$\text{Error sum of squares} = \sum_{i=1}^{N} X_i^2 - \frac{1}{N}\left[\sum_{i=1}^{N} X_i\right]^2$$

Using this criterion, the original N groups (objects) are reduced to $N-1$, $N-2$, and so on until all N objects are in one group. The method is monotonic, strictly Euclidean, and is especially suitable for continuous attributes, although it can be used with mixed data types. It is often used in vegetation analysis (e.g. Kent and Wathern 1980).

Fisher (1969) proposed an algorithm which differs from that of Ward by trying some sub-optimal fusions at specified places in the hierarchy in the hope that one of them may lead to a better clustering than that obtained by optimizing the hierarchical route.

The Lance and Williams combinatorial model

For some strategies, the information needed to calculate object/group and group/group measures is contained in the original dissimilarity matrix, so that once that is calculated the original data are no longer needed. Such measures are said to be combinatorial (Lance and Williams 1967a). There are certain measures for which combinatorial solutions either do not exist or have not been found, and the original data are required throughout the clustering. Examples are the Manhattan metric, for which no centroid solution exists (in any case, centroid clustering is now considered obsolete), and information statistic measures, which use information gain as a dissimilarity measure (Williams 1971).

The linear combinatorial model of Lance and Williams (1967a) is as follows. Suppose we have two objects (or groups) i and j, with a distance (or dissimilarity) d_{ij}, and these objects (or groups) fuse to form a new group k. We have a third object (or group) h for which the distances d_{ih} and d_{jh} are known, and we require the distance d_{kh}. The model is

$$d_{kh} = \alpha_i d_{ih} + \alpha_j d_{jh} + \beta d_{ij} + \gamma |d_{ih} - d_{jh}|$$

The values of $\alpha_i, \alpha_j, \beta$ and γ define the particular strategy in use, γ may take the value of 0 or ± 0.5, the other parameters are normally functions of the numbers of objects in the groups concerned. Values for the commonly used strategies are as follows:

(i)   Single linkage $\alpha_i = \alpha_j = 0.5, \beta = 0, \gamma = -0.5$

(ii)   Furthest neighbour $\alpha_i = \alpha_j = 0.5, \beta = 0, \gamma = 0.5$

(iii)   Centroid (UPGMC) $\alpha_i = n_i/n_k, \quad \alpha_j = n_j/n_k, \quad \beta = -\alpha_i \alpha_j,$ $\gamma = 0$, and n_i is the number in group i, n_j is the number in group j, $n_k = n_i + n_j$. The combinatorial formula is valid only for squared Euclidean distances.

(iv)   Median (WPGMC) $\alpha_i = \alpha_j = 0.5, \beta = -0.25, \gamma = 0$. The combinatorial formula is valid only for squared Euclidean distances.

(v)   Group average (UPGMA) $\alpha_i = n_i/n_k, \alpha_j = n_j/n_k, \beta = \gamma = 0$.

(vi)   Simple average (WPGMA) $\alpha_i = \alpha_j = 0.5, \beta = \gamma = 0$.

(vii)   Ward's error sum-of-squares $\alpha_i = (n_h + n_i)/(n_h + n_k),$ $\alpha_j = (n_h + n_j)/(n_h + n_k),$ $\beta = -n_h/(n_h + n_k), \gamma = 0$.

(Lance and Williams 1967a; Mather 1976). The only strategies for which γ is non-zero are nearest and furthest neighbour, and Lance and Williams (1967a) stated for the remaining strategies (with $\gamma = 0$) the string of measures associated with successive fusions will be monotonic, i.e. the dendrogram will contain no backward links, if $\alpha_i + \alpha_j + \beta \geqslant 1$.

Using their combinatorial formula, Lance and Williams (1967a) developed a flexible clustering strategy in which $\alpha_i = \alpha_j$, $-1 \leqslant \beta \leqslant 1$, $\alpha_i + \alpha_j + \beta = 1$, $\gamma = 0$, that is, it is completely defined by the cluster intensity coefficient β. The method can be made to range from space-dilating ($\beta = -1$) to space-contracting ($\beta = 1$). However, there is some danger in adjusting a parameter until one obtains a result which is pleasing (but it may be useful in dissection). Booth (1978) found the flexible clustering strategy with $\beta = -0.25$ to be useful in the analysis of tree distribution data. This value of β is found to be suitable for many purposes (Clifford and Stephenson 1975).

The combinatorial formula has some deficiencies connected with what happens in the case of equal smallest dissimilarities. Sibson (1971) has shown that the arbitrary choice of a pair leads to failure of the algorithm to produce a well-defined cluster method in the majority of cases. Furthermore, the formula cannot immediately be extended to deal with situations more complicated than uniting groups two at a time. Nevertheless, the approach can be useful (Jardine and Sibson, 1971).

Wishart (1969c) gave an algorithm for implementing the combinatorial model, and noted that for Ward's method it appeared to be considerably more efficient than the algorithm outlined by Ward. The general usefulness of the combinatorial model should not obscure the fact that more efficient algorithms are available for some of the individual methods, e.g. single-linkage (Sibson 1973) and complete-linkage (Defays 1977). For a specified number of groups, the partition produced by the furthest neighbour version of the combinatorial algorithm can be regarded as an approximation to the partition for which the maximum dissimilarity between a pair of objects in a given group is minimized. Such a partition may be obtained by an algorithm of Hansen and Delattre (1978). The optimal partitions into different numbers of groups can overlap, and the furthest neighbour algorithm cannot be guaranteed to find them (Gordon 1981).

An aesthetic objection to the distance approach is that the distance matrix is just a middle step between the actual data and the final clustering structure. It would be better to have algorithms and models that directly connect the data and the desired clustering structure (Hartigan 1975).

The interpretation of dendrograms and comparison of methods

The end-point of a hierarchical strategy is a dendrogram. However, the problems are not over once a dendrogram has been obtained, as the interpretation of dendrograms needs some care. A dendrogram is a two-dimensional representation, and does not contain all of the information residing in the dissimilarity or similarity matrix from which it was derived.

In a dendrogram, the objects will be set out at the bottom of the diagram with the root at the top. A scale at the side gives the values for the measure at which successive fusions occurred. The order in which the individuals are listed at the bottom of the dendrogram needs to be considered. On computer printouts, the order usually has no special significance, it is simply a computational convenience. Each node represents a fusion, but the order of the points within a fusion can be changed without altering the sequence of fusions. Therefore, the order of the points is not an indication of intergroup similarity or difference.

In a study of the responses, measured on 6 occasions, of wheat cultivars to stem rust, Thompson and Rees (1979) used the Canberra metric with group average clustering (UPGMA). The resulting groups were identified on a principal co-ordinate ordination chart, on which the distribution of points representing the cultivars was strongly curved. The groups obtained contained cultivars with similar degrees of susceptibility, and the sequence of fusions in the dendrogram showed which cultivars most resembled which other cultivars in rusting characteristics.

The intergroup relationships can be found by calculating the group/group measures between all pairs of terminal groups and then performing a principal co-ordinate analysis.

It has been suggested that a large change of level in a dendrogram indicates the 'correct' number of groups. However, Everitt (1974) showed that such a procedure may be misleading. A large change in level in a dendrogram is a necessary but not a sufficient condition for the presence of clear-cut clusters. Indeed, Clifford (in Williams 1976) pointed out that a line purporting to represent a level of heterogeneity must never be drawn across a dendrogram unless the fusion strategy in use is strictly space-conserving, and since the only completely space-conserving strategy is the obsolete centroid strategy, such a line should not be used at all. Indeed, it would not be logical to use a hierarchical method in this way because in such methods the groups through which the process passes are not necessarily optimal in themselves; the best route may be obtained at the expense of some slight reduction in homogeneity of the individual groups. If a particular number of groups is required, a non-hierarchical method would be preferable.

Different clustering strategies explore different aspects of a data set. It may therefore be useful to be able to compare dendrograms. Jardine and Sibson (1971) distinguished three types of purpose for which we might wish to compare two or more classifications, using the term in its broad rather than statistical sense: (i) measurement of the congruence of classifications, e.g. whether a classification based on one battery of attributes is congruent with one based on another battery; (ii) testing independence, such as whether a classification based on soil type is independent of one based on plant species found on the soils; (iii) assessing dependence, e.g. the degree to which a classification based on one battery of attributes can be predicted from another. Rohlf (1974c) reviewed such methods, chiefly of type (i).

Dendrograms have two fundamental properties, their structure and their group compositions. If the interest is mainly in group compositions, a comparison may be made by noting the number of pairs of objects associated in the same group in one dendrogram but not in the other, and vice versa, then summing the two numbers. The smaller the result, the more similar are the dendrograms in group composition. However, if the main interest is in the relationships between groups, then some measure of dendrogram structure is needed. One of the simplest methods is to determine the number of nodes by which given pairs of objects are separated in two dendrograms. A measure of dissimilarity between dendrograms is obtained by subtracting the larger number from the smaller in each comparison and summing over all pairs of objects (Clifford, in Williams 1976).

Sneath (1977a,b) described a significance test for the distinctness of a pair of clusters derived from a dichotomy in a dendrogram in certain conditions, and Sneath (1979) gave a BASIC program for applying the test.

A dendrogram implies a new set of distances between objects which can be found from the distance level of the lowest link joining the two objects in the dendrogram. These distances are ultrametric. A necessary and sufficient condition for an observed dissimilarity coefficient to be exactly represented by a dendrogram is that it should satisfy the ultrametric inequality (Chatfield and Collins 1980). Sokal and Rohlf (1962) defined the cophenetic correlation coefficient as the product-moment correlation coefficient between the elements of the original dissimilarity matrix and the dissimilarities implied by the dendrogram. Other measures have been proposed (e.g. Hartigan 1967; Jardine and Sibson 1968; Phipps 1971; Williams and Clifford 1971; Duncan and Estabrook 1977).

Rohlf and Fisher (1968) attempted to answer the question of how large a cophenetic correlation coefficient should be in order to show that a dendrogram provides a reasonable summary of the observed dissimilarities. Their study suggested that a value of 0.8 is needed for rejecting the null hypothesis of a single cluster versus the alternative hypothesis of a hierarchical system of clusters. However, Rohlf (1970) warned that a value near to 0.9 does not guarantee that a dendrogram provides an adequate summary of the observed dissimilarities. The advantages and disadvantages of the cophenetic correlation coefficient were discussed by Sneath (1966), Williams and Clifford (1971), Sneath and Sokal (1973), and Rohlf (1974c).

Farris (1969) compared various hierarchical clustering methods using the cophenetic correlation coefficient and found that of the methods discussed above the group average method (UPGMA) produced the greater coefficient values. However, even greater values could be obtained from dendrograms with backward links. Because of this the usefulness of the cophenetic correlation must be doubted. The appropriateness of the cophenetic correlation coefficient as a criterion depends on many factors. For example, Farris concluded that it is not a direct measure of the degree to which a classification describes the distribution of attribute states, and that if it is desired to use the term 'optimal' for classifications in which the most similar objects are clustered together, then the cophenetic correlation coefficient should not be employed as a criterion. Also, an elongated cluster might be better represented by single-linkage than by average-linkage clustering, even though the latter might have a larger cophenetic correlation coefficient (Sneath 1969).

Apart from the cophenetic correlation coefficient, dendrograms can be compared with the original distances using a normalized Minkowski metric (Sokal 1977) or stress measures (Gower 1975b; Orloci 1978). Johnson (1967) preferred rank correlation measures, which assume only that the observed dissimilarities have only ordinal significance. Huber (1974) suggested the use of the Goodman–Kruskal gamma statistic with the single-linkage and complete-linkage methods.

Muller (1975) compared the application of several clustering methods to seven sets of data comprising samples from trivariate normal distributions. He found that Wishart's mode analysis produced much better results than did the single-linkage method, but was inferior to Ward's method and to methods based explicitly on the assumption that the clusters are multivariate normally distributed (Chapter 14). UPGMA performed poorly in this test.

Cunningham and Ogilvie (1972) used two measures to compare the methods discussed above (except flexible clustering) in clustering artificial data. UPGMA was found to be always at least as good as any other method, and WPGMA and furthest neighbour were almost as good. The performance of the nearest neighbour method depended on the type of data used, but was considered to be generally poor. Ward's method was found to be prone to chaining and it and the centroid and median methods introduced distortion into a data set which was designed to represent a pure hierarchical structure. On the other hand, UPGMA did not perform well on the data of Muller (1975), and Ward's method did rather better in that comparison. Hawkins *et al.* (1982) commented that in the comparison made by Cunningham and Ogilvie, UPGMA outperformed the other methods mainly on data sets which had a fair degree of inherent chaining, and could not be regarded as coming from a mixture of normal parent distributions. Also, there were only 20 objects.

Other comparative studies were performed by Gower (1967a), Boyce (1969), Pritchard and Anderson (1970), Kuiper and Fisher (1975), Baker and Hubert (1975), and Milligan (1981). Hartigan (1977) suggested that the choice of a correlation coefficient may influence the results of such comparisons, e.g. in the study of Baker and Hubert. Orloci (1978) discussed some measures in detail. He commented that methods based on the similarity levels at which fusions occur may be too restrictive because they give too much weight to quantitative differences among dendrograms which may be quite similar in their fusion topologies. He thought that there was much to be said for the method of Phipps (1971) which describes dendrograms by their fusion topologies. Similarly, Farris (1973) thought that a single measure would not provide a reliable comparison of dendrograms as similarities in clusters at lower levels would swamp marked dissimilarities in clusters at higher levels. He favoured examination of the extent to which clusters of one dendrogram are fragmented among clusters of the other. Orloci (1978) noted that this method not only detects distortion but also identifies the clusters subjected to distortion.

Jardine and Sibson (1971 p86) discussed the question of continuity, i.e. the output dissimilarities should be continuous functions of the inputs. They regarded discontinuity as being crippling because the effects of computational rounding errors and of statistical sampling errors are not predictable in practice and can lead to completely misleading results. Only the single-linkage method satisfied this, and other conditions. However, Gower (1975b) thought that their rejection of all other hierarchical methods was too extreme, and that some of their criteria are not essential (see also Williams, Lance *et al.* 1971). In practice, the only real criterion against which a clustering method can be judged is its usefulness, and a method which is useful for one set of data may not be for another set, or for a different objective. Hence, the main need is to understand the properties of the various methods, and studies on artificial data with known properties are extremely useful for this purpose.

The single-linkage method gives solutions which are invariant in structure under a monotone transformation of the dissimilarity measure, so it is good for data having only ordinal significance (Chatfield and Collins 1980).

Everitt (1979) reviewed several empirical studies in which the single-linkage method did less well, on average, than other methods, notably Ward's method. Everitt (1974) described an empirical investigation of the properties of the single-linkage, centroid (UPGMC), and Ward's methods, using Pythagorean distances, when applied to artificially-generated bivariate data containing: (i) one circular group; (ii) two circular groups; (iii) two elongated groups. With (i) the single-linkage method suggested that there was no real structure in the data. The dendrogram from the centroid method suggested the presence of two main groups with several unallocated points. Ward's method also suggested the presence of two groups. With (ii), all three methods recovered their original structure. With (iii), the single-linkage method gave two major clusters which were essentially the groups generated. The centroid method gave two groups which were not those generated. Ward's method suggested the presence of two, or possibly three, groups which were not those generated.

The above study illustrates the fact that different methods make different assumptions about the structure of the data and tend to impose a structure on the data. There is no clear evidence in many studies that the investigator distinguishes between the properties of the different clustering methods. However, used thoughtfully the heuristic alternation of descriptive and proscriptive approaches can yield satisfying insights (Sokal 1977).

One way of assessing the performance of clustering techniques, and of comparing different techniques, is to simulate populations of known structure and then find to what extent the techniques recover that structure (e.g. Everitt 1980 p77; Gardner 1978).

As random populations are characterized by a certain degree of clustering, van Groenewoud (1983) defined cluster analysis as the detection and identification of groups of objects that resemble each other more than they resemble members of other groups, to a greater degree than can be expected from random distribution. Van Groenewoud applied single-linkage and minimum sum of squares methods to two sets of data generated to have a known random distribution. With 50 species (attributes) the sum of squares method showed strong clustering, whereas the single-linkage method showed no clear structure.

Even if it is not implicit in the method, many clustering methods have minimum-variance properties. Among 13 such methods, Wishart (1969b) listed: Sorensen's (1948) complete-linkage, furthest neighbour, centroid, simple average (WPGMA), and group average (UPGMA) methods. The concept of minimum within-cluster variance has great appeal for practical applications, because such a grouping will have a high predictive value. Such methods are especially useful in dissection. However, care is needed if they are used in an exploratory analysis, as they may divide dense clusters in an unacceptable manner, as was shown above and by Wishart (1969b). Many natural objects have distributions in attribute space which are quite different from this model, and dendrograms obtained from such methods would represent the relationships rather poorly (Sokal 1977). Minimum variance methods naturally tend to produce hyperspherical clusters, but the data may not have this type of structure. Jancey (1974) described a modified single-linkage clustering method which tests the possibility of dividing points into r groups of minimum size s on the basis of discontinuities.

Fisher and van Ness (1971) approached the problem of selecting a 'best' clustering procedure via decision theory, which tells us to restrict our attention to admissible decision rules. They listed nine admissibility conditions, and specified which of these were satisfied by the following five clustering methods (the number given with the method indicates the number of conditions it failed to satisfy): (a) single-linkage, 1; (b) furthest neighbour, 2; (c) minimum least squares – k fixed, 4 with one condition not applicable; (d) hill climb least squares – k fixed, 5 with two not applicable; (e) centroid, 5.

Here again, it is clear that the single- and complete-linkage methods satisfy the greater number of conditions, indeed these methods failed only on the convex admissibility, which Fisher and van Ness admitted does not seem universal since it eliminates many reasonable clusterings. It should be noted that when Fisher and van Ness found that an algorithm was inadmissible, it will not necessarily fail on all data sets.

Fisher and van Ness (1971) also noted that if two admissible clustering schemes give different dendrograms, one might wonder whether the data were suitable for a tree structure. This would seem to be a legitimate use of the flexible clustering strategy of Lance and Williams (1967a). Thoughtful use of the properties of different clustering methods can reveal certain properties of the data. For example, the single-linkage method maximizes the minimum intercluster distance at each step. The furthest neighbour method minimizes the maximum cluster diameter at each step. If different dendrograms result from the use of the two methods, then both of the above objectives cannot be attained at the same time. A comparison of the two trees would be revealing. Since, at present, there is little knowledge of how to choose between many different methods of calculating similarity coefficients and hierarchical clustering algorithms, presentation of the data under various methods would give some, admittedly non-quantitative, information on the reliability of any dendrogram obtained. A more concise method of presentation would be to run several methods and give the diameter of the set of dendrograms obtained. This would help avoid the computation time objection to Hartigan's (1967) approach, but not the need for a metric.

Ling (1972) described a technique called SHADE which is applied to a similarity or distance matrix before clustering, and to the rearranged matrix after clustering. It shows graphically if any distinctive clusters have been produced. Methods for studying the internal cohesiveness of clusters can be useful, e.g. normal probability plots of within-cluster interpoint distances (Cohen *et al.* 1977).

Mukkattu (1974) clustered vegetation data using a hierarchical agglomerative sum of squares method. Five groups were selected, and their values for soil phosphorus, potassium, and calcium were compared by profile analysis (Morrison 1967). For a given group, the profile is the vector of group means of the variates. The method assumes that the within-group covariance matrices are homogeneous and non-singular. The null hypothesis, that the type profiles are identical, was rejected. The next hypothesis, that the profiles are parallel, was accepted. The closeness of the parallel profiles was then tested by a simple analysis of variance on the group totals. With two groups this is a simple two-tail t test.

Baum (1978) applied various clustering methods to genera of grasses. The results were compared by finding the correlations between the attributes and the various classifications (treated as attributes). The measure is in terms of Shannon's information, and resulted in D (distance) values from 0 to 1 for each attribute. The 'best' classification is that with the lowest overall score. The classifications were also compared using Gower's (1971b) rotational technique.

Gower (1971b) suggested that two sets of distances, such as ultrametric distances derived from dendrograms (e.g. Banfield 1976), can be compared by first computing their respective principal co-ordinate values. The two sets of co-ordinates are then superimposed so that their centroids match, and one set is rotated, and possibly reflected, until the best match possible is obtained. The sum of the squared distances between the positions of the objects on the two sets of co-ordinates is used as a measure of discordance, called R^2. This technique is called classical Procrustes analysis. Matrices of R^2 values may themselves be treated as distances and may be examined by principal co-ordinate analysis (Gower 1971b; Williams, Dale and Lance 1971; Krzanowski 1971). This is a form of metric multidimensional scaling.

The above approach is similar to the INDSCAL (individual differences scaling) method in which each individual constellation of points is represented relative to the axes of an overall analysis (Carroll and Chang 1970; Carroll 1972).

Gower (1975a) described generalized Procrustes analysis for more than two sets of distances. He noted that the method was not intended to rival INDSCAL, but to provide a complementary analysis giving different information. The results can be summarized in the form of an analysis of variance, although the degrees of freedom were not known and there was no sampling theory for the observed sums of squares.

Rohlf (1974c) thought that such methods might provide useful insights. Classifications based on different batteries of attributes yield similar, but not identical, classifications. If there is a continuum of different classifications, then the variability might have to be treated as a form of 'experimental error'. On the other hand, if there are batteries of attributes which would yield clusters of classifications, then a number of special purpose classifications would be indicated, and each would summarize a different set of attributes.

If the classifications being compared are for a given set of objects using different batteries of attributes, or if the effect of adding attributes is being studied, then it is essential that the clustering method used should be insensitive to small changes in the data. Otherwise the researcher cannot know if the differences observed between classifications are due to differences in the data, or result simply from instability in the clustering method. The single-linkage method is the only hierarchical method which satisfies both the stability requirement and certain other simple requirements about the way in which a classification should represent the mutual dissimilarities of objects (N. Jardine in discussion in Cormack 1971).

Two sets of clusters, or a set of clusters and one of *a priori* groups, may be compared through a contingency table. The correlation between the two can be tested by chi-square or an information measure, i.e. we can test the hypothesis that one set has no predictive value for the properties of the other set (e.g. Feoli 1976; Orloci 1978 p251).

The particular distributions of plant species present special problems in cluster analysis as they do in ordination. Robertson (1979) discussed the properties of three hierarchical clustering methods applied to simulated coenoplanes.

A practical example of hierarchical cluster analysis

As an example of hierarchical cluster analysis we can use the component values obtained from the soil data in Chapter 5. First we need a half-matrix of interobject dissimilarities. As the components are orthogonal and the space is Euclidean we can use Pythagorean distance. The soils formed some reasonably distinct groups in the first two dimensions (Figure 5.7), but these were less clear in a scatterplot on the first and third axes (Figure 5.8). Also, the first four components accounted for 87% of the total variance. It is therefore of interest to do a cluster analysis on two, three, and four components.

The first step is to compute a half-matrix of Pythagorean distances. For an exploratory cluster analysis the next step is to compute a single-linkage cluster analysis. Figure 13.2 gives the single-linkage dendrogram obtained by the algorithm of Ross (1969d) from a half-matrix of Pythagorean distances among objects (soils) on the first two components. As in Figure 5.7, the podzols (soils 19–24) form a distinct group at a small distance level (0.174). The base-deficient brown earths (soils 1–6), the base-rich brown earths (soils 7–12), the brown podzolic soils (13–18), and the peaty gleys (37–42) form a group at a slightly larger distance level (0.349).

At increasing distance levels the string of points representing the peaty podzols in Figure 5.7 (soils 25–30) join with the larger group, initially because of the proximity of points 29 and 39. Similarly, the string of points representing the gleys (soils 43–48) in Figure 5.7 join with the larger group, initially because of the proximity of points 48 and 6. Of the hill peats (soils 31–36), soils 31 and 32 join at a low level (0.116) but at the same level soil 34 joins with 26 (a peaty podzol). All of the hill peats form a group at distance level 0.465, but this group also includes two peaty podzols (26 and 30).

Figure 13.3 shows the dendrogram for a group average (UPGMA) clustering of the same distance half-matrix. It is immediately evident that the distances in this dendrogram are larger than those in the single-linkage dendrogram, which illustrates the different group size dependence of the two methods. As in the single-linkage dendrogram, a large group of 25

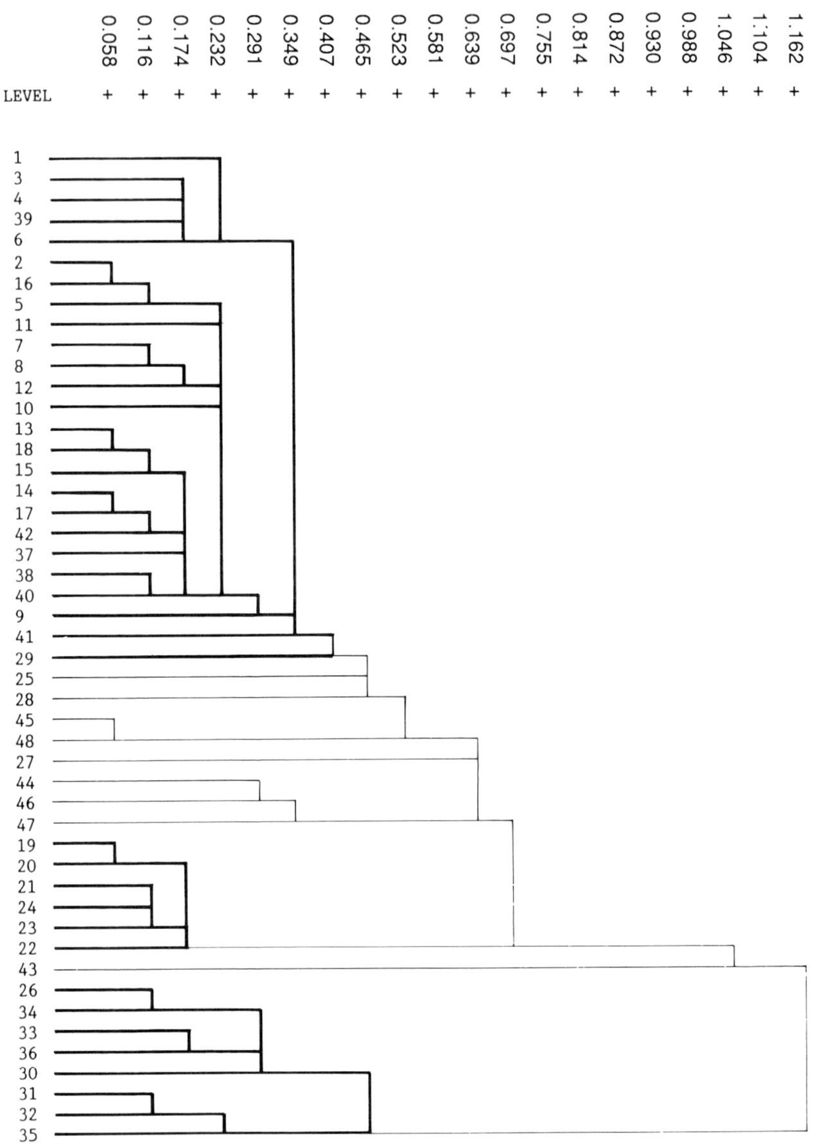

Figure 13.2 Single-linkage dendrogram obtained from Pythagorean distances computed from two components of the soil data in Chapter 5

soils forms at a moderately-low level in the dendrogram but with the difference that in this case soil 29 joins this group at a much earlier stage. Here, the podzols form a group in a very similar way to that formed in the single-linkage dendrogram, as do the hill peats plus peaty podzols 26 and 30.

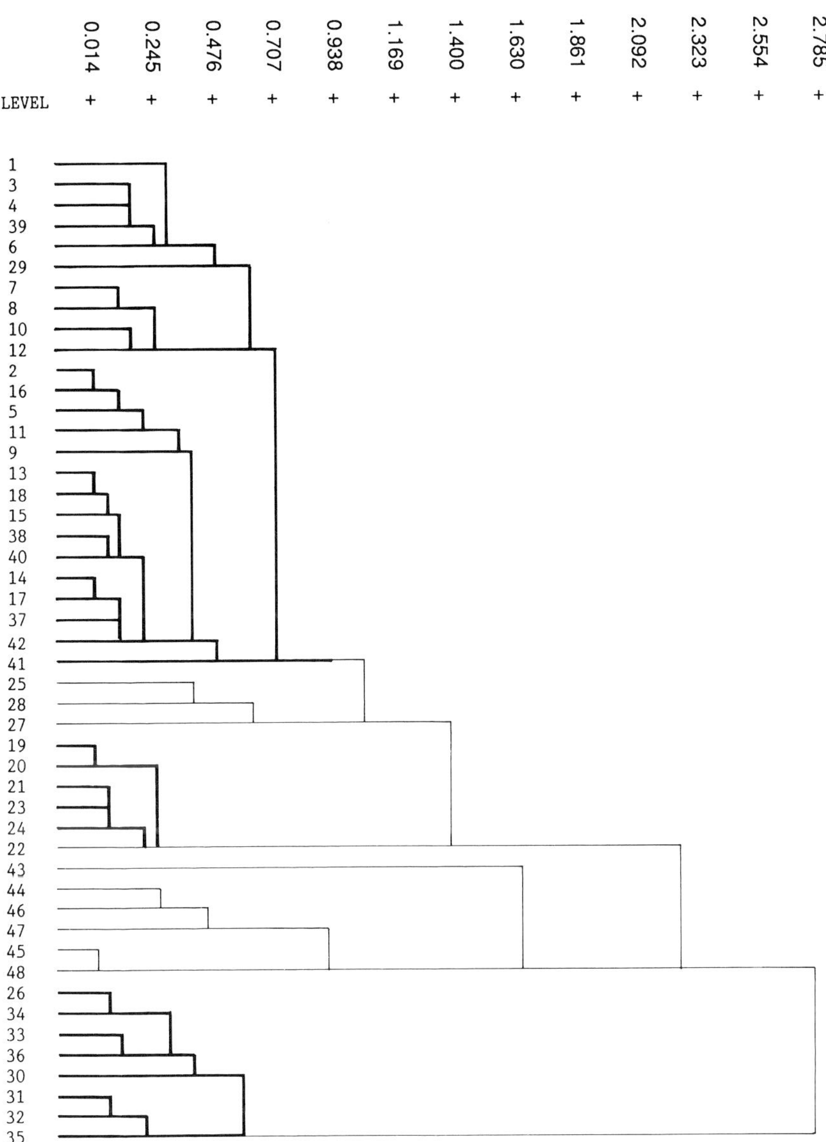

Figure 13.3 Dendrogram for a group average (UPGMA) clustering of the distances used in Figure 13.2

Figure 13.4 shows the single-linkage dendrogram obtained from a half-matrix of distances computed using three components. The distance levels are noticeably larger than those in Figure 13.2. This is not due simply to the mathematical effect of adding another component, as the distances are divided by the number of dimensions to give average distances. The

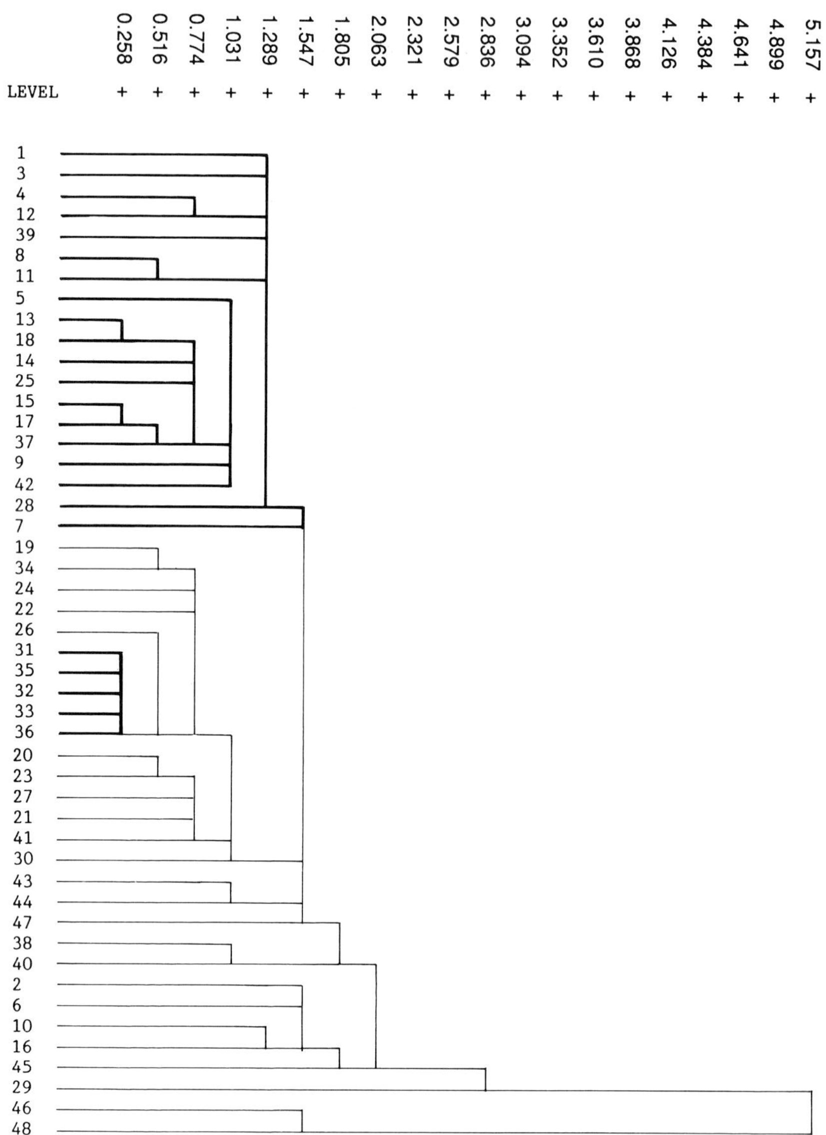

Figure 13.4 Single-linkage dendrogram obtained from Pythagorean distances computed from three components of the soil data used in Figure 13.2

difference is due to the different interpoint relationships in three dimensions. The pattern of fusion of the points is also different. For example, the podzols (soils 19–24) no longer form a clearly-defined group. Five of the hill peats cluster together at the lowest level (0.258), more so than in Figure 13.2, but soil 34 joins them with other soils at a slightly greater level (0.774).

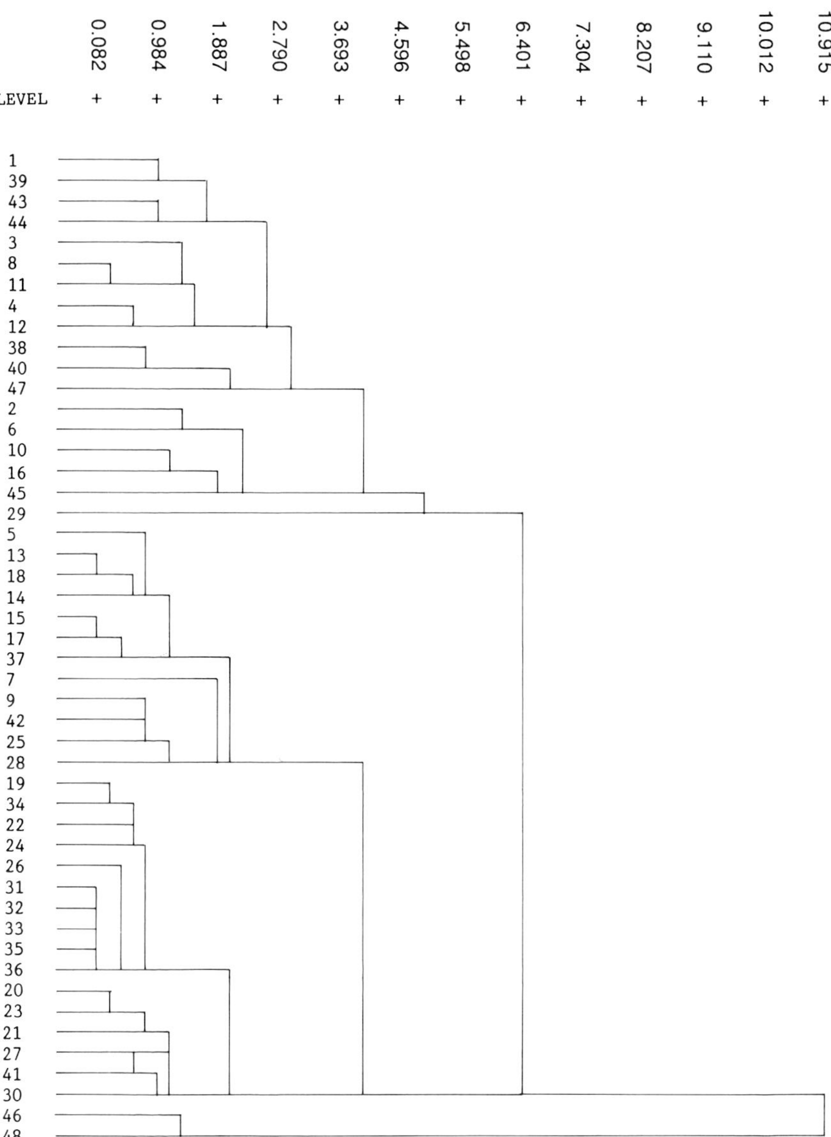

Figure 13.5 Dendrogram for a group average (UPGMA) clustering of the distances used in Figure 13.4

Many of the points that formed the large group in Figure 13.2, with the addition of soils 25 and 28, also form a group in Figure 13.4. However, two of the base-deficient brown earths (soils 2 and 6), one of the base-rich brown earths (soil 10), one of the brown podzolic soils (soil 16), and half of the peaty gleys (soils 38, 40 and 41) are elsewhere in the dendrogram.

Within this large group the interrelationships of the points have changed to some extent.

The UPGMA dendrogram for three dimensions (Figure 13.5) is also different from that in two dimensions, much of the original group structure having changed.

The minimum spanning tree used to compute the single-linkage dendrogram in Figure 13.2 is given in Chapter 15 (Tables 15.1 and 15.2).

CHAPTER 14

SOME OTHER CLUSTERING METHODS

A variety of clustering methods has been proposed. Most have not been widely used because they are mere variants of already-established methods, or they have undesirable properties, or they are impracticable in programming terms. These methods may be found in the text-books already cited. A few of the methods are discussed here.

Non-hierarchical methods

Various methods have been proposed which seek to minimize some function of the root mean square pairwise dissimilarity within elements of a partition and maximize the root mean square pairwise dissimilarity between members of different elements of the partition. They are sometimes called sum of squares or variance methods (Jardine and Sibson 1971; Sneath and Sokal 1973; Clifford and Stephenson 1975). The properties of these methods are not well known, but Jardine and Sibson (1971) noted that there is not, in general, a unique partition on which the measure is optimized.

One example of such a method is k-means clustering, for which an efficient algorithm was given by Hartigan and Wong (1979). In this method, the user specifies the number of clusters required and the initial cluster centres. The latter may be obtained by any method which suits the objectives of the user, for example using some hypothesis about the data, or from a hierarchical method. The first step in k-means clustering is to allocate each object to its closest cluster centre. The means of the clusters are then calculated and are taken to be the new cluster centres. The observations are then checked in turn to see if a move to a different cluster results in a decrease in the overall sum of squares. When an acceptable partition has been obtained, the object nearest to each cluster centre can be found and nominated as the 'type object' of that cluster. The overall sums of squares provide a means of comparing the results of k-means clustering with partitions obtained in other ways (e.g. Howard and Howard 1981, 1988). This method may not find the best grouping (global optimum) but it does

197

find one that could not be improved by moving any single observation to another cluster (local optimum). Such algorithms are called transfer algorithms.

Sparks (1973) drew attention to the importance of the choice of initial cluster centres, although it is usually not clear how this choice is to be made. Hartigan (1975) and Hartigan and Wong (1979) suggested methods for selecting initial cluster centres. However, care is needed using such methods because they may give cluster centres which do not give a local optimum. A simple example of this is the test example in Hartigan (1975 p 88). If this set of data is processed using initial cluster centres chosen by the method of Hartigan and Wong (1979), a different configuration is obtained for the same number of groups. A solution to this problem is to use more than one method for selecting initial cluster centres, which could include the centroids of clusters obtained by some other method, possibly hierarchical, as initial cluster centres. k-means and similar methods are applicable only to strictly Euclidean systems and it should be pointed out that the results obtained with different numbers of clusters are not necessarily hierarchical.

k-means clustering is probably the most generally useful of the non-hierarchical methods. Its practical application can be illustrated in the analysis of 14 physico-chemical attributes of 0 to 5 cm depth soil samples collected from 48 woods in and around the English Lake District. A principal component analysis showed that four eigenvalues were greater than unity and together accounted for 78% of the variation. The seven largest eigenvalues accounted for 92% of the variation. A single-linkage cluster analysis showed that there were no discontinuities which would enable a classification to be made. However, if it were necessary to produce a dissection, e.g. as a stratification for future sampling, then k-means clustering would be well-suited to such a purpose.

The method of Hartigan and Wong (1979) was applied to the first seven components of the soil data using initial cluster centres obtained by methods suggested by Hartigan (1975) and Hartigan and Wong (1979). Table 14.1 gives the overall sums of squares for different values of k (number of groups), together with the mean square ratios R. For a given value of k

$$R = \left[\frac{\text{overall sum of squares for } k \text{ groups}}{\text{overall sum of squares for } k + 1 \text{ groups}} - 1 \right] (M - k - 1)$$

where M is the number of points in the analysis. As a crude rule-of-thumb, an overall mean square ratio greater than 10 for a given value of k would justify the use of $k + 1$ groups, especially if the value of R for $k + 1$ is smaller than that for k (Hartigan 1975).

In Table 14.1 the large overall sum of squares for 8 groups using the second method for selecting initial cluster centres does not follow the

Table 14.1 Overall sums of squares and mean square ratios (R) for different k-means partitions of the 48 soils using the first seven components and two different methods for selecting initial cluster centres

k	Overall ssqs	R	Overall ssqs	R
3	335.16	6.73	336.51	6.96
4	290.70	9.40	290.53	5.12
5	238.56	3.63	259.64	7.46
6	219.56	5.09	220.46	6.12
7	195.31	5.55	191.83	5.41
8	171.52	2.85	221.85	11.24
9	159.84	—	172.22	5.85
10	NC		149.26	2.13
11	NC		141.15	4.32
12	NC		126.02	5.79
13	NC		108.13	—

NC No initial cluster centres obtained with this method

general trend of decreasing values as k increases, and probably represents a sub-optimal configuration of points. Hence, the R value of 11.24 can be ignored. The first method for selecting initial cluster centres failed for 10 to 13 groups. This method computes the co-ordinates of possible cluster centres from a quick initial grouping, and for some combinations of k and M it does not obtain the required number of cluster centres. The second method for selecting initial cluster centres selects actual points as initial cluster centres. Table 14.1 illustrates the value of using at least two methods for selecting initial cluster centres, values from both methods can be mixed. In this example, for each method the R values do not suggest that any particular value of k is preferable, so the user could choose a value appropriate to the objectives of the study.

Transfer algorithms can be used to optimize other criteria, for example to maximize the number of agreements with binary class-predictors, as in Gower's (1974) maximal predictive classification.

The ISODATA method (see e.g. Gnanadesikan 1977) also requires the user to specify the number of clusters required and the initial cluster centres. A splitting and a lumping criterion are also required. The method has three stages. In the first, or sorting, stage each of the N objects is assigned to one of the initial clusters, according to the Pythagorean distance of the observation from each cluster centre. When all N objects have been allocated, any empty clusters are discarded and the remaining clusters are replaced by their mean vectors. In the second stage, three statistics are calculated: (i) for each cluster, the average distance of the points from the cluster centroid; (ii) for each cluster, the covariance matrix; (iii) across all clusters, the average intracluster distance. These three statistics provide measures of the tightness of the clusters.

In the third stage, new clusters are formed by splitting apart, or lumping together, existing clusters. The user has the choice of two methods of splitting. In the first method, if the largest diagonal element of a cluster covariance matrix, i.e. the maximum variance of one of the co-ordinates, is greater than the given splitting parameter, that cluster is split along the co-ordinate with the largest variance. In the second method, if the largest eigenvalue of a cluster covariance matrix is greater than the splitting criterion then the cluster is split along the direction of the first principal component. Two clusters are fused if the distance between their centroids is smaller than the lumping parameter, and a new centroid is calculated.

The three stages are iterated. The number of clusters initially specified may not be critical, as it is not necessarily achieved. Provided that adequate iterations are allowed, the method appears to be reasonably robust. The method is, of course, scale-dependent by virtue of using variances and Pythagorean distances.

The additive property of sums of squares means that the 'classification efficiency' of a partition of objects into k groups can be expressed by the within- and between-group sums of squares. Information measures can be treated similarly but average similarity cannot (Orloci 1978 p239). To avoid circularity, and to examine the general usefulness of the groups, comparisons should be made on attributes not used in the clustering procedure. A formal analysis of variance may be possible if the assumptions are satisfied. If the assumption that the populations from which the samples were obtained are multivariate normal can be justified, various hypotheses about the k groups can be tested (Orloci 1978 p242). However, such an assumption is likely to be difficult to justify in practice.

Anderberg (1973 p204) discussed the advantages and disadvantages of methods for measuring the similarity between partitions (mutually exclusive clusters).

Most clustering methods are based on a dissimilarity or similarity measure. The models often contain implicit assumptions about the shape of the clusters, and the necessity for storing the dissimilarity or similarity matrix results in problem with computer memory for large numbers of objects. Attempting to overcome such problems, Gitman and Levine (1970) presented on algorithm for partitioning objects into unimodal 'fuzzy sets'. A fuzzy set is characterized by a membership function which associates with each object a number between 0 and 1 which represents the 'grade of membership' of the object for the set. Advantages claimed for this method are that its demands on computer memory and time are not large, and the shape of the distribution of points in a group is not restricted (see also Bezdeck 1974; Zadeh 1977). The method does not seem to have been much used in biology.

Divisive methods

Edwards and Cavalli-Sforza (1965) suggested that points might be divided into sets such that the sum of squares of distances between sets is a maximum (i.e. it is a polythetic divisive technique). Gower (1967a) drew attention to the collossal computational labour involved in the direct examination of all the possible partitions of N points. On a computer with $5\,\mu s$ access time it would take 100 hours for $N = 21$ and $54\,000$ years for $N = 41$. Orloci (1967a) devised a criterion for overcoming this heavy computational load, but it is not monotonic and reversals in its value can occur (Sneath and Sokal 1973).

Gower (1967a) raised the question of whether the method should maximize intergroup sums of squares of the distances between group centroids. The sums of squares method takes into account the sample size of each cluster, and since some samples of equal importance in the overall classification will in certain cases be based on greatly unequal numbers of objects, the method based on maximizing distance between centroids may be preferable. However, the centroid method has the disadvantage that there may be points in one cluster which are nearer to the centroid of another cluster. With well-separated clusters, the maximum sums of squares and maximum distance between centroid methods will yield the same results, but so will most other methods. The real test of a method lies in its ability to deal with more challenging cases (Sneath and Sokal 1973).

Other methods include association analysis and related methods. These divisive methods were devised primarily for the classification of objects described by binary attributes. The well-known monothetic technique of association analysis, due to Williams and Lambert (1960), was originally intended for use with plant species presence/absence data in terrestrial ecological survey. The method divides a set of objects into two subgroups based on the two states of a single character chosen to maximize chi-square. The subsets are similarly divided, and the process ends when a predetermined number of groups is reached, or when the measure of homogeneity as expressed by chi-square has fallen below a critical level. Geometrically, association analysis can be represented in attribute space as finding the binary attribute (normalized by standard error) for which the distance between the centroids of the two groups defined by it is maximal (Gower 1967a; Sneath and Sokal 1973). Gower (1967a) showed that the method contained implicit weighting for abundance.

The method, though widely used, had two disadvantages. Firstly, it was unduly sensitive to highly-skewed data, i.e. to the presence of rare species or the absence of common ones. Secondly, if the decision function itself was used as a level for drawing a dendrogram, reversals were frequent (Williams 1976). Other authors (e.g. see Jardine and Sibson 1971) determined the fit of each partition in terms of the information loss induced by the partition. Lance and Williams (1968) have adapted the association analysis method to the information statistic 2I (see also Williams and Lance 1977). So-called polythetic analyses seek that bisection which minimizes the information loss or some other related function regardless of whether bisection corresponds to the range of the states of any of the selected binary attributes (Jardine and Sibson 1971), see, for example, Macnaughton-Smith *et al.* (1964) and Macnaughton-Smith (1965).

Methods of monothetic association analysis have been used in ecology, but Jardine and Sibson (1971) stated that the available methods are unsatisfactory in several respects: (1) they are ill-defined, data can be readily constructed for which no bisection induces a unique minimum information loss; (2) partition into two subsets at each stage is an arbitrary choice; (3) their application in taxonomy is restricted to discrete-state attributes which do not vary within populations. They further noted that the use of these methods appears to rest upon a confusion between classification and diagnosis. Monothetic association analysis produces a hierarchical classification by choosing a diagnostic key, based on the available attributes, which is in a precise sense optimal, but the production of optimal diagnostic keys is not the primary purpose of classification in ecology or taxonomy.

Gower (1967a) discussed the properties of association analysis and noted that the distance between a pair of quadrats, one of which contains a rare species or does not contain a common species, will be exaggerated compared with distances between 'typical' quadrats. This seems to be an undesirable property, since the occurrence of a rare species may be a matter of chance, whereas the absence of a common species may be due to a real difference in ecological conditions.

Macnaughton-Smith's (1965) dissimilarity analysis is a divisive method for binary data, using an information statistic. An ALGOL program for its implementation was given by Tipper (1979), who listed examples of its use in geology. Its use in plant ecology was described by Macnaughton-Smith *et al.* (1964), who noted that a modification of the method could be used for quantitative attributes (see also Massart and Kaufman 1983).

Williams and Dale (1965) pointed out that certain clustering methods and combinations of methods, e.g. association analysis, the method of Macnaughton-Smith *et al.* (1964), and some methods using Mahalanobis' D^2, may result in locally Euclidean spaces separated by discontinuities or embedded in non-Euclidean spaces.

Crawford and Wishart (1967, 1968) proposed an alternative to association analysis for the assessment of a large binary ecological matrix. It assumes that ecological groups are determined by species which occur frequently with high density, rather than those which are frequent but isolated or are frequent in rich areas. Latent structure analysis (Lazarsfeld and Henry 1968) offers a more direct approach to finding clusters within which attributes are uncorrelated. The clusters found are disjoint, but objects are given probabilities of belonging to them rather than being definitely assigned to one (Baker 1962).

Methods not based on a dissimilarity measure

Indicator species analysis (Hill *et al.* 1975) is a divisive polythetic method which was developed to solve a specific problem in the analysis of vegetation data. The problem was to divide a sample of stands of vegetation into groups which summarize the range of variation in the vegetation and have a straightforward ecological interpretation. The method should produce a key which would enable new stands to be identified with ease, and should be capable of handling a large number of stands. The method begins by ordering points representing the objects (stands) on the first axis of a reciprocal averaging ordination (Chapter 9). Then, the points are divided into two groups at the centre of gravity. Next, the five species whose occurrences are most nearly confined to stands on one or other side of the division are identified ('indicator species') and the division point is adjusted so as to maximize the 'indicator efficiency' of the indicator species in subsequent allocation of additional stands to the appropriate groups. Each subgroup so obtained is subjected to the same procedure of reciprocal averaging and division, and the process is iterated to give as many groups as the user requires. At each division, each indicator species is given a score of 1 or -1 according to the branch of the dichotomy that it indicates. Thus, a key for the identification of new stands is obtained.

One problem which occurs in using indicator species analysis with vegetation data is that the method depends on the presence of at least one indicator species, and is likely to be reliable only if two or more are present. High species diversity reduces the probability that this condition will be met. Also, any additional stand to be identified should have the same dimensions as those used in the original survey, should be selected according to the same criteria, and should be searched thoroughly for the indicator species, otherwise misclassification may occur without being recognized as such (Hall and Swaine 1976). Another problem is that reciprocal averaging does not preserve ecological distances.

Bunce *et al.* (1975) extended the application of indicator species analysis to data obtained from maps, using grid squares as objects. However, the extension of the use of indicator species analysis beyond its original application to become a general clustering method is open to objections. The use of the first axis of a reciprocal averaging ordination is based on the phytosociological theory of floristic gradients, which may or may not be true. The idea is that the first reciprocal averaging axis represents a floristic gradient and thus contains all the important information. With other types of data the concept of a floristic gradient will not be valid and the first reciprocal averaging axis will not necessarily contain all the important information. A major weakness of reciprocal averaging is that if more than one major ecological axis exists, the spurious polynomial axes responsible for the 'arch' effect interact with the ecological axes and confound interpretation (Austin 1985). Gauch and Wentworth (1976) found that reciprocal averaging usually gave poor results with environmental data.

The effectiveness of the dichotomous division of the first axis about its centre of gravity will depend on how well the information contained in the data can be represented in a single ordination axis, and how uniformly the objects are distributed along the axis. This is unknown in advance, and is likely to vary considerably for different sets of data. Hence, the properties of groups produced by this method will be largely unpredictable. Another problem is that reciprocal averaging is a special application of correspondence analysis, which was developed to handle contingency tables. It is not well suited to continuous variables. Although it is sometimes suggested that they should be converted to qualitative attributes, this involves some loss of information and distortion (see Section 3.1 and Chapter 9).

Howard and Howard (1981) compared an ISA clustering with a k-means clustering of component values extracted from the same data. The results showed that the ISA classes tended to be diffuse and heterogeneous to varying degrees. It is difficult to find an application for which such group properties are acceptable. Adamson (1984) found that Major Soil Groups predicted from ISA land classes were in some cases greatly different from those obtained from soil maps. Groups are more generally useful if they have minimum variance properties.

Moss (1985) showed that groupings obtained from map data using two-way indicator species analysis (TWINSPAN) were, in general, less good than those produced by k-means clustering. Some of the tests concerned comparisons based on predictions of plant and bird species distributions, data not used to form the clusters.

The only really satisfactory way of handling mixed continuous and discrete data types appears to be via Gower's (1971a) general similarity coefficient. Used with principal co-ordinate analysis (q.v.) this produces an ordination which is strictly Euclidean if there are no missing values. k-means clustering can then be performed on the ordination scores of the objects. Howard and Howard (1988) applied this combination of methods to mixed data obtained from maps, and compared the overall sums of squares of the k-means partitions with those of previously-obtained partitions from indicator species analysis. The results supported the previous conclusion that indicator species analysis tends to produce heterogeneous groups, although the authors did cite one example in which the two methods produced similar results. k-means clustering consistently produces groups which have minimum-variance properties, which is valuable for most practical applications. If the number of objects is sufficiently large to cause problems with computer memory space when computing the general similarity coefficient, Lefkovitch's (1976) transformation method may be used (see Chapter 7). However, with a very large number of objects it may be statistically more desirable to divide the objects into subsets and analyse them separately, thus testing the stability of the analysis.

One special group of methods is based on the assumption that the data are a mixture of multivariate normal distributions. Several of these methods were reviewed by Holgersson and Jorner (1978) and Hawkins *et al.* (1982). Friedman and Rubin (1967) proposed a method based on minimizing the generalized variance within the groups (i.e. the determinant of the pooled within-groups sums of squares and products matrix). This idea seems attractive at first sight, because it is equivalent to minimizing Wilks' criterion. However, it was criticised by Marriott (1974), who pointed out that if the data consist of samples from a mixture of unimodal distributions, the groups defined by this procedure will be the truncated centres of these dispersions mixed with the tails of other distributions. The dispersion matrix estimated within groups will not be an estimate of the dispersion matrix of the underlying distributions even if these are identical, and there is no reason to expect that it will be the same within the artificial groups found by the clustering process.

Marriott (1971) attempted to overcome the difficulties experienced by other workers using the generalized variance approach by assuming a uniform distribution as a null hypothesis. If the null hypothesis is true, the effect of an optimum subdivision on the generalized variance can be predicted. If subdivision of data into *g* groups reduces the generalized variance by much more than that predicted, it is reasonable to suppose that it corresponds to an inherent grouping in the data. The method appears to work reasonably well, although when the modes are near together and the distributions overlap considerably, separation may be impossible even for very large samples. On the other hand, some peculiar unimodal distributions of an extremely leptokurtic type may be subdivided. It is necessary to test each pair of groups in isolation to see whether they should be recombined. Advantages of this method are: (1) great flexibility in the data that can be handled and in the use of concomitant observations; (2) independence of scale and of linear transformations. Its disadvantages are: (1) the mathematical basis is not altogether solid, in particular the criterion for subdivision is rather arbitrary; (2) a significance test, though theoretically possible, does not yet exist; (3) the computational load is heavy (Marriott, 1974).

Beale (1969) noted that attempting to minimize the sum of squares of the deviations of the observations from their respective cluster centres is equivalent to maximum-likelihood if all clusters are assumed to be (hyperspheroidally) normally distributed with a common variance. Two methods based on a maximum-likelihood approach have been suggested. Marriott (1974) considered the method of Day (1969) to be almost the only classification technique that is entirely satisfactory from the mathematical point of view. It assumes a well-defined mathematical model (that the underlying distributions are multivariate normal with equal dispersion matrices), investigates it by well-established statistical techniques, and provides a test of significance of the results. The fact that it is difficult to apply, and in many situations is unrealistic reflects the complexity of the question that cluster analysis is trying to answer.

The other maximum-likelihood method is that of Scott and Symons (1971). Their approach was to make maximum-likelihood estimates of the means, variances and covariances, and the identifying parameters that assigned the sample points to the groups. The resulting estimates indicated that the identifying parameters should be chosen to minimize the generalized variance. Marriott (1974) considered their conclusions to be misleading, and could not justify the method of minimizing the generalized variance used in this way. However, he also pointed out (Marriott 1975) that the assumption of underlying normal distributions with equal dispersion matrices is seldom strictly true in practice, and in many practical situations, when the proportions in the underlying distributions are approximately equal, minimizing the generalized variances gives a sensible and reasonably robust clustering procedure, although it is better regarded as a heuristic approach rather than an estimation process applied to a particular model. Marriott (1975) showed that Day's approach leads to consistent estimators, while that of Scott and Symons does not.

Marriott (1971) noted that the smallest increase in the generalized variance occurs when an individual is added to the group for which Mahalanobis' generalized distance of the individual from the group mean is minimum. This suggests that once the cores of the groups are known, allocation by multiple discriminant or canonical variate analysis could be used to the same effect, provided that the theoretical requirements of these methods are satisfied.

Everitt (1974) described an empirical investigation of the properties of some clustering methods applied to artificially-generated bivariate data containing: (i) one circular group; (ii) two circular groups; (iii) two elongated groups; (iv) one circular and one elongated group. The trace $(\mathbf{W})$ criterion and the $|\mathbf{W}|$ criterion recovered the groups generated in (ii), but the $g^2|\mathbf{W}|$ criterion, suggested by Marriott (1971), failed as it preferred a single-group solution until the means of the two groups were separated by a certain minimum value. With (iii) the trace $(\mathbf{W})$ criterion, which assumes that the groups are all approximately hyperspherical, divided each elongated group and assigned half of each group to one cluster and half to the other, i.e. it imposed on the data the nearest approach possible to circular clusters. On the other hand, the $|\mathbf{W}|$ criterion, which assumes that all groups present have the same shape but are not necessarily spherical, recovered the groups as generated. The $g^2|\mathbf{W}|$ criterion failed with these data. With (iv), the $|\mathbf{W}|$ criterion included some points from the elongated group in the cluster containing the circular group. The $g^2|\mathbf{W}|$ criterion suggested a 5-group solution. A criterion suggested by Scott and Symons (1971), which does not assume that the groups have the same shape, recovered the original structure.

Apart from the problems caused by the fact that the criteria assume that the groups have certain characteristics which are unlikely to be known in advance, the methods using the trace (**W**) and the |**W**| criteria require large amounts of computer time even for moderate-sized data matrices.

Lee (1979) described a criterion for testing the hypothesis that the observations are a random sample from a single multinormal population versus the alternative that the observations are from two multinormal populations with different means. The criterion is equivalent to the likelihood ratio criterion but is more manageable computationally.

Katz and Rohlf (1973) proposed a method called function-point cluster analysis as the logical extension to their function-plane technique for oblique rotation in factor analysis (q.v.). The local maxima of their function correspond to local regions of high density of points, i.e. clusters. Each cluster comprises all data points which are 'under' the same peak in the surface defined by the function. Clusters obtained at different hierarchical levels are not necessarily nested.

Adaptive methods

Most clustering methods are non-adaptive, that is, the algorithm proceeds towards a solution by means of a fixed clustering method which may, to a greater or lesser extent, impose a structure on the data. However, an ideal clustering method would be adaptive. It would make an initial exploration of the data to find the types of clusters that are probably present, and would then modify the clustering algorithm to suit whatever structure is considered to be most likely. Some methods which attempt this were discussed by Sneath and Sokal (1973). Two in particular will be noted here, both based on the single-linkage method.

A number of clustering methods possess variance constraints; Wishart (1969b) discussed 13. Implicit in the minimum variance approach is the concept that clusters should have no significant overall variance or spread, and this implies that in the case of a unimodal swarm the distribution would be split into an arbitrary number of compact sections. Forgey (1964, 1965) argued that clusters should correspond to data modes, and there can only be as many classes as there are distinct modes. No variance constraint is implied, or should be induced, for when a mode is elongated rather than spherical, the distribution merely reflects some internal factor of variation for the corresponding class. Forgey interpreted a data mode as a continuous dense swarm of points separated from other modes by either empty space or a scattering of 'noise' data. The 'noise' data may result from sampling errors or they may be interpreted as those natural phenomena associated with the intersecting tails of disjoint continuous distributions. The cluster analysis problem is therefore to isolate the dense centres irrespective of the interference (Wishart 1969a, b).

Wishart (1969a, b) took the single-linkage method as a basis for his 'mode analysis'. The satisfactory mathematical properties of the single-linkage method have already been outlined. Wishart's solution to the clustering problem was to remove the 'noise' data, to cluster the remaining dense swarms by single linkage, and then to reallocate each noise datum according to a similarity criterion. This was achieved by selecting a distance threshold r and a density limit k. From each object the method tests whether k or more objects lie within r, if so, the object is considered 'dense' (this corresponds to counting the number of links to the object in a single-linkage clustering). The 'dense' points are then clustered by a single-linkage method at the threshold r, and the resultant clusters delimit the dense cluster nuclei. Each 'non-dense' point is then allocated to a cluster by some criterion. By fixing k and varying r a hierarchical classification is produced. A severe decision demand is placed on the user in selecting r and k.

Marriott (1971, 1974) concluded that the method of Wishart was the best available for detecting and identifying a natural grouping, and it is unlikely to produce a meaningless or misleading answer unless there is a mixture of continuous and binary attributes. However, he also pointed out that it is insensitive in detecting elongated modes, and the choice of the value of k may affect the conclusions. He (1971) made the following points: (1) The search for modes by dense points can lead to misleading results when continuous distributions are involved unless samples are very large; (2) The dense points are defined in terms of a 'spherical' scanning device. This has certain advantages; discrete distributions can be included and there is no problem of unwanted classification on the basis of a single variate. On the other hand, the method is scale-dependent, is rather sensitive to the inclusion of highly-correlated variates, and the existence of genuine multimodality can be masked by the inclusion of irrelevant variates, especially if the modes are ellipsoidal rather than spherical.

Everitt (1974) applied mode analysis, using CLUSTAN 1B, to artificially-generated bivariate data containing: (i) a single circular group; (ii) two circular groups; (iii) two elongated groups; (iv) one circular and one elongated group. For (i), solutions of from two to ten groups were given, the other data gave solutions of from two to 17 groups. For all sets of data, the groups as generated could be obtained from the lower order solutions because of the prior knowledge of the real structure. Without such knowledge, interpretation is difficult. Wishart has suggested that the solution with the largest number of clusters should be considered to be the most significant, but Everitt's tests do not support this.

A similar strategy was proposed by Shepherd and Willmott (1968). In this strategy, the data are clustered in two stages. The purpose of stage one is to determine which objects are most likely to be at or near the centres of groups. In this stage, a single-linkage method is used, followed by a process of discarding peripheral objects until only compact nuclei remain. The severity of this reduction process is determined by a group reduction criterion. This results in a series of cluster nuclei which are fed to stage two, in which the cluster nuclei are expanded using a modified pair-group average linkage method. A readmission criterion determines how easy readmission into a group should be. An advantage of this technique for cluster expansion is that the clusters can have skewed distributions.

Although the basic idea is attractive, Shepherd and Wilmott's single-linkage cluster analysis method was criticised by Gower and Ross (1969), who also commented that ill-defined concepts of multiple linkage were used to provide information on structure which is very simply derived from the minimum spanning tree. Two problems in using single-linkage cluster analysis to locate data modes are that (a) the dense modes may only be clearly identifiable if the sample is very large, and (b) small local aggregations of points may occur outside the modes, and may act as false cluster centres. For example, points which are randomly scattered in a two-dimensional plane, e.g. by a planar Poisson process, show some signs of clustering on visual inspection. Conversely, points generated by a 'hard-core' model so as not to be random do, in fact, appear to an observer to be random (Chatfield and Collins 1980). A formal test of the hypothesis that the points are randomly scattered over the plane is useful (see Ripley 1977).

Rohlf (1970) proposed some methods which could adapt to possible trends of variation found within clusters as they are being formed. A non-linear version allowed the isolation and description of clusters of various shapes, including parabolic and ring-shaped.

Extrinsic methods

Sometimes, the nature of the problem suggests that certain groupings of points are not logical or helpful. An example of this is in pollen analysis, where cluster analysis can be used to delimit pollen zones. Clearly, a sample from a given depth in the profile cannot belong to the same pollen zone as a sample from higher or lower in the profile if samples from intermediate depths belong to a different zone or zones. In this case, clustering requires the constraint that, if an agglomerative procedure is being used, amalgamation can occur only if the elements are continuous, i.e. if the levels at which the samples were collected are adjacent, or the groups of levels are not separated by any level which is not in either group.

Gordon and Birks (1972) used a constrained single-linkage procedure, with a choice of simple Manhattan or a Canberra metric, using proportions of pollen type, for this purpose. They drew attention to the fact that with this procedure, if many extra samples are taken close to suspected zone boundaries, it is conceivable that dissimilarity coefficients between successive levels near the zone boundary could be of the same order of magnitude as those in other parts of the sequence. This would result in the zone boundaries not being clearly defined. Because of this, they also tried two constrained divisive procedures, one based on information content and one based on the sum of squared deviations, the latter using three different weighting methods. The results showed that similar stratigraphical divisions were obtained with all methods, and the resulting zonations corresponded to those arrived at independently by visual inspection. Minor differences were apparent, particularly in the exact location of zone boundaries. The results were discussed in detail, and the limitations and potentialities of the procedures were considered.

In another, similar, study on data from two other sites, Birks (1974) found a striking consistency between the zonation schemes suggested by the various numerical analyses, and also a close agreement between those zones and the original zonation schemes proposed independently by the original investigators although, again, there were minor differences in detail.

A similar problem occurs in soil mapping, where a multiplicity of small areas makes a map difficult to use. Webster and Burrough (1972b) proposed a method for smoothing local irregularities and eliminating small fragmentary parcels while recognizing widely-separated, but otherwise similar, parcels of soil in the same class. This was done by varying the contribution made to the similarity index by location according to the distance between pairs of sites.

For a further discussion of constrained clustering, see Gordon (1980, 1981).

Some classification methods supplement multiple regression methods. A model is defined for one set of attributes while the searching criterion is based on another set. Macnaughton-Smith (1963) and Hunt *et al.* (1966) gave methods in which a single attribute is predicted from a multivariate set of predictors. Mantel and Valand (1970) suggested a method for extending this approach to multivariate prediction. Beeston and Dale (1975) showed that Macnaughton-Smith's (1965) multiple predictive analysis, which arose from an extension of extrinsic classification, can be useful for the preliminary evaluation of experimental programmes. It is, in fact, the analogue for discontinuous data of the analysis of covariance (Williams 1976 p131).

Here we might mention automatic interaction detection (Sonquist and Morgan 1963; Kass 1975; Hawkins and Kass 1982), which is appropriate when one wishes to predict a dependent attribute from a set of predictors which are all discrete (nominal or ordinal). The method works through a number of stages in each of which the categories of one of the predictors are investigated and are grouped into classes by merging those categories which do not differ significantly in the values for the dependent attribute. The method is related to Anova and to regression.

Overlapping methods

An example of sequential agglomerative hierarchical overlapping methods is Jardine and Sibson's (1968) Bk family, which is based on modifications of the single-linkage method which allow overlaps of one or more objects per cluster at any given level. Although the resulting hierarchy resembles the original dissimilarity matrix more as more objects are allowed to overlap at any given level, it becomes increasingly more difficult to draw and interpret the dendrogram (Sneath and Sokal 1973; see also Cole and Wishart 1970; Jardine and Sibson 1971; Rohlf 1974a,b).

Jardine and Sibson (1971) discussed cluster methods which, by generalizing to allow stratified systems of overlapping clusters, succeed in avoiding the defects of the methods outlined in the previous section, and also recover more information than does the single-linkage method, although this is achieved at the cost of greater complexity in the resulting classification. Their two sequences of stratified cluster methods, Bk and Cu, were shown to be satisfactory within Jardine and Sibson's axiomatic framework. Bk operates by a restriction on the size of the permitted overlap between clusters, Cu operates by a restriction on the diameter of the permitted overlap between clusters, which is proportional to the level of the cluster. Bk is likely, in certain circumstances, to be more unstable under extension of range than is Cu. Whenever there are well-marked groups with intermediates, Cu is likely to produce clusterings which are more stable as the range is extended, because it is less vulnerable to alteration in the number of objects intermediate between clusters. Cu pays for this greater stability by requiring stronger assumptions about the significance of the underlying dissimilarity coefficient than does Bk.

The algorithm of Jardine and Sibson (1968) was improved by Cole and Wishart (1970), but it can only deal with up to 60 objects.

CHAPTER 15

GRAPHS AND TREES

Introduction and definitions

Graph theory has proved to be of value in solving practical problems, or in showing for which problems no solution is possible (e.g. see Burns 1983).

In the graph theoretical sense, a graph is a set of points called *vertices*, and of relations between pairs of vertices which are indicated by lines called *edges*. A graph is said to be *connected* if every pair of distinct vertices is joined by at least one chain, which need not be direct. A *maximally-connected* graph has edges between all pairs of vertices. A *minimally-connected* graph contains only one direct or indirect path between every pair of vertices. Removal of one edge from a minimally-connected graph disconnects it into two subgraphs. A *maximal complete subgraph* is a set of vertices for which all the edges are present and which is not contained within a larger, totally-linked set.

A graph is said to be a *tree* if it is connected and has no circuits. The removal of any one edge of a tree yields a disconnected graph, as the edge removed constituted the unique chain joining two vertices. Hence, a tree is a minimally-connected graph. If all vertices of graph G are included in tree T, then T is said to *span* G. In graph theory, an edge implies only that a connection exists between two vertices. However, it is possible to associate with an edge a real number which can be called its *length*. A *minimum spanning tree* has the smallest possible sum of the lengths of the vertices.

A special family of graphs is the family of *directed* graphs, also known as *networks*, which imply direction in the edges. A directed tree has edges with direction and a unique path from one vertex, called the *root* of the tree, to all other vertices. A conventional dendrogram is an example of such a graph.

Some uses of graphs and trees

From the above, it is easy to see that a set of objects may be represented by vertices and their dissimilarities by the lengths of the edges. Hence, by breaking the graph at various special levels based on the lengths of the edges one can form clusters of vertices (objects) which are more similar to each other than the chosen level.

The utility of the graph theoretical approach in this context is threefold. First, graphs serve as illustrative devices that enable many investigators to understand a variety of problems connected with cluster analysis. Second, the graph theoretical approach enables us to derive certain properties of clusters from well-established theorems of graph theory and to employ graph-theoretical tools as solutions to specific problems, Third, they provide extra information when superimposed on ordinations (Sneath and Sokal 1973).

Some cluster analyses leave invariant the dissimilarities between certain pairs of objects. The set of elements left invariant by the single-linkage method corresponds to the edges of the minimum spanning tree. Efficient algorithms are known for finding minimum spanning trees, and this is the most efficient computational procedure for single-linkage cluster analysis (Gower and Ross 1969; Ross 1969b, c, d). New points can be added to an existing minimum spanning tree (Roger 1971; Roger and Carpenter 1971), and the method can be used to handle sets of data which are too large for conventional classification procedures (Ross 1969a).

It is useful to superimpose a minimum spanning tree on an ordination in order to provide an additional perspective of interpoint relationships (e.g. Gower and Ross 1969; Rohlf 1970; Schnell 1970).

To illustrate the nature of a minimum spanning tree, Table 15.1 shows the minimum spanning tree from which the single-linkage dendrogram in Figure 13.2 was derived. The table lists the objects from 2 to N, the object to which each is joined in the tree, and the distance. To make it easier for the user to interpret the tree, and to draw it as a graph, it is shown in Table 15.2 with the links sorted. The tables were produced using the algorithms of Ross (1969b, c). In Figure 15.1, object number 1 is found together with the object to which it is joined (in this case number 10), and the distance. Then, in a new column, object number 10 is printed together with the object to which it, in turn, is connected (in this case 12), together with the distance, and so on. In that way, a chain of connections is built up, in this case of objects 1, 10, 12, 7, and so on to object 26. At that point the links cease, so the algorithm goes back along the chain to look for side chains. In this example it finds one from object 33 to 31, 32, and 35. When this side chain is printed, the number of the first point (33) is printed in the same column as in the previous chain. Other chains are found and printed in the same manner.

Table 15.1 The minimum spanning tree from which the single-linkage dendrogram in Figure 13.2 was derived

Object no.	Joins no.	Distance
2	5	0.1087
3	4	0.1556
4	39	0.1708
5	11	0.2090
6	3	0.1643
7	12	0.1687
8	7	0.1137
9	37	0.2694
10	1	0.2991
11	8	0.2242
12	10	0.1782
13	42	0.1253
14	42	0.1051
15	18	0.0753
16	2	0.0552
17	14	0.0440
18	13	0.0433
19	20	0.0566
20	41	0.6952
21	24	0.0710
22	21	0.1621
23	20	0.1486
24	23	0.0693
25	28	0.4116
26	33	0.2719
27	28	0.5996
28	40	0.4146
29	39	0.3659
30	27	1.1178
31	33	0.4358
32	31	0.0670
33	36	0.1521
34	26	0.0685
35	32	0.1816
36	30	0.2696
37	14	0.1162
38	40	0.0758
39	1	0.2189
40	15	0.1182
41	37	0.2973
42	5	0.1780
43	44	1.0033
44	46	0.2868
45	48	0.0137
46	47	0.3406
47	45	0.5983
48	6	0.4722

Length = 11.7881

```
 1   10   0.2991
    10   12   0.1782
       12    7   0.1687
          7    8   0.1137
             8   11   0.2242
               11    5   0.2090
                  5   42   0.1780
                    42   13   0.1253
                       13   18   0.0433
                          18   15   0.0753
                             15   40   0.1182
                                40   28   0.4146
                                   28   25   0.4116
                                   28   27   0.5996
                                      27   30   1.1178
                                         30   36   0.2696
                                            36   33   0.1521
                                               33   26   0.2719
                                                  26   34   0.0685
                                               33   31   0.4358
                                                  31   32   0.0670
                                                     32   35   0.1816
                                40   38   0.0758
                    42   14   0.1051
                       14   17   0.0440
                       14   37   0.1162
                          37   41   0.2973
                             41   20   0.6952
                                20   23   0.1486
                                   23   24   0.0693
                                      24   21   0.0710
                                         21   22   0.1621
                                20   19   0.0566
                          37    9   0.2694
                  5    2   0.1087
                     2   16   0.0552
 1   39   0.2189
    39    4   0.1708
       4    3   0.1556
          3    6   0.1643
             6   48   0.4722
                48   45   0.0137
                   45   47   0.5983
                      47   46   0.3406
                         46   44   0.2868
                            44   43   1.0033
    39   29   0.3659


Length = 11.7881
```

Figure 15.1   The minimum spanning tree of Table 15.1 printed with sorted links

Although minimum spanning trees have proved to be useful in interpreting relationships among objects, any well-defined graph structure could be used depending on the subject matter of the analysis and the objectives of the user (Dale, in Williams 1976 p103). For theoretical reasons, taxonomists may be interested in minimum spanning trees, Wagner trees, Hu trees (Farris *et al.* 1970), or other types, but other researchers do not need to be interested only in the same range of possibilities (cf. Wallace and Boulton 1968; Gower 1974).

The graphical aspects of several sequential agglomerative hierarchical non-overlapping clustering methods are illustrated in Sneath and Sokal (1973). The concept of connectivity of graphs is related to the structure of clusters. If all members of a cluster are so close together that the corresponding subgraph is fully connected, the cluster will be compact. Maximal connectivity is a criterion for complete-linkage cluster analysis. Hansen and Delattre (1978) showed that the complete-linkage clustering problem can be reduced to the problem of optimally colouring a sequence of graphs, and proposed an efficient algorithm to solve it. Minimally-connected sets of vertices are likely to give long strung-out clusters, and minimal connectivity is the criterion for single-linkage cluster analysis. Matula (1977) described standard connectivity properties of graphs which were shown to characterize several clustering methods. He noted an intrinsic shortcoming of the complete-linkage method which makes it suspect for general application.

Various indices of connectivity have been proposed, for example see Gabriel and Sokal (1969) and Sokal and Crovello (1970).

Jardine and Sibson's (1971) Bk clustering has a graph-theoretical representation (Rohlf 1974a). As the value of k increases from unity (minimum spanning tree), the graphs become more highly connected, and they always contain a subset of edges representing distances to the k nearest neighbours of each point. At a given level, edges are removed if they are greater than the threshold value. Then, maximal complete subgraphs are sought, and are compared pairwise. If a pair contains more than $k - 1$ common members they are fused to form a cluster. The process is repeated until no further fusion occurs (see example in Gordon 1981). These graphs are much simpler to represent than are the corresponding dendrograms. Indeed, they can be presented as a kind of two-dimensional ordination (Sneath and Sokal 1973).

Gordon (1981) discussed constrained clustering procedures in terms of graph theory, and noted that constrained clustering is equivalent to searching for a cut-set of the graph, that is the set of edges whose removal would cause the graph to break down into separate subgraphs. In some cases, every edge is a cut-edge.

Wirth *et al.* (1966) defined an acceptable cluster as follows: (i) A collection of specimens (objects) is isolated for some fixed value of similarity if each specimen in the collection is less similar to every specimen outside the collection than that similarly; (ii) An acceptable cluster is a collection of specimens (objects) which is isolated for some fixed similarity value but which contains no smaller cluster isolated for the same fixed value. Hence, members of an acceptable cluster show a discontinuity with non-members, and an acceptable cluster cannot be further subdivided without the parts being less isolated than the whole.

They presented a computer method for finding such clusters, based on graph theory (see also Estabrook 1966). The values of the similarities between pairs of objects were assigned to the edges, and all pairs of objects (vertices) at least as similar as some given similarity value were linked. Such linked collections of objects constituted acceptable clusters. The method is based on the partition of the collection of objects into equivalence classes (clusters) which are maximal connected subgraphs. Clusters were described not only by the similarity value associated with them, but also by the numerical expression of the isolation of a cluster, called its 'moat'.

Wirth *et al.* (1966) defined the moat of a cluster as the difference between the level at which it was formed and the level at which it is incorporated into a larger cluster. It can be thought of as a measure of the empty space around a cluster. Obviously, the sizes of the moats obtained from different methods will reflect the criteria of formation of the clusters. For a given method, the sizes of the moats may be informative but as the hierarchical structure may distort the real relationships this measure should be used with caution (Gordon 1981).

The 'connectedness' of a cluster may be defined as

$$\frac{\text{Total no. of existing connections} - \text{no. of necessary connections}}{\text{Total possible connections} - \text{no. of necessary connections}}$$

(Wirth *et al.* 1966; Clifford and Stephenson 1975).

In any cluster of N objects, the minimum number of connections required for the cluster to exist is $N - 1$. The maximum number of possible connections is $N(N - 1)/2$. Hence, the connectedness is given by

$$\frac{e - (N - 1)}{N(N - 1)/2 - (N - 1)}$$

where e is the existing number of connections (edges). By definition, a complete-linkage group will have a value of unity (maximally interconnected), while a single-linkage group could take the minimum value of zero.

Wirth *et al.* (1966) found that when this method was applied to 31 members of the Oncidiinae (Orchidaceae) the hierarchy showed good separation of clusters. The results were considered to be more satisfactory than those of other cluster methods using the same data. In some cases, clusters were linked by 'articulation points', i.e. objects intermediate between two clusters. When such objects were removed, the graph split into subgraphs which remained isolated over a wide range of levels. The graph theory model provided a theoretical framework within which the nature of the relationships could be examined.

In studying the taxonomy of the Liliatae, Clifford and Williams (1980) found that a network-generating program which, unlike minimum spanning tree programs, considered both first and second nearest neighbours, proved useful in studying the affinities and helped to account for difficulties in classification.

Graphs and trees have uses other than in cluster analysis. For example, Romane *et al.* (1977) found that the minimum spanning tree of a set of species of phytoecological relevés was easier to interpret than the results of a correspondence analysis, and emphasized especially the importance of geomorphological and edaphic attributes.

Dale (1977a) called two plant species positively associated if they occurred together in a significantly greater number of cases than would be expected if they were distributed randomly, and negatively associated if they occurred together in significantly fewer cases. Such associations could arise if the species had similar, or very different, ecological requirements respectively. Dale discussed the problem of detecting association between species in terms of directed graphs, and (Dale 1977b) applied the theory to data from a mixed forest in Ontario. The analysis produced a graphical representation of the phytosociological structure of the plant community, and its graph-theoretical properties could be used to test theories about the nature of the community. In this particular community, Goodall's (1966b) hypothesis was supported. Goodall (1966b) described a plant community as a group of species tending to be found in the same location because of the special features of the latter, together with other species which are not necessarily favoured by the non-living environment but rather by the presence of one or more species of the first group.

Fitch (1977) discussed the problem of finding the most parsimonious unrooted tree, which is a minimally connected graph containing t branch tips, $t - 2$ strictly bifurcating ancestors, and $2t - 3$ edges. This problem was related to the interpretation of nucleotide sequences.

CHAPTER 16

CHOOSING A STRATEGY

General considerations

It is impossible to give hard-and-fast rules for selecting a pattern analysis strategy because so much depends on the researcher's objectives. Virtually any data set will contain some form of pattern, and possibly more than one form. Pattern is always pattern-for-an-agent, the result of an interaction between the properties of the data and the mind of the researcher studying them (MacKay 1969). Therefore we cannot talk about pattern being true or false (if the analysis has been done correctly), only about it being profitable or unprofitable. The researcher's objectives are important because they define what is, or is not, profitable.

Tukey (1962, 1969) suggested that data analysis is essentially an open-ended, interactive, and iterative procedure in which the researcher takes different approaches to illuminate different aspects of the data. Allen *et al.* (1984) emphasized that biological systems contain information at various scales and no single view can encompass all of it. Gradients which appear in the data at one scale may become curved or confused at a different, but valid, scale of observation. A variety of techniques, each with its particular scaling properties, may be necessary in order to characterize the relevant behaviour of the system. This view is supported by the finding of Chardy *et al.* (1976) that a combination of several ordination techniques yielded a more complete description of variation in benthic fauna than did any single method.

Generally, pattern analysis methods are used to 'fish' for insights which might be used, for example, to generate hypotheses from survey type data. Such hypotheses could be tested by experiments followed by more formal

221

statistical procedures. Some pattern analysis methods can be used to test hypotheses informally, and certain methods can be used in both formal and informal ways.

Thus, statistical tests can be used with principal component analysis if the data are substantially multivariate normal. However, the lack of multivariate normality does not invalidate all aspects of the method, which can be used in an informal way assuming only that the data are measured an interval or ratio scales and the attributes are linearly related.

Because they can be used in a non-parametric way, some pattern analysis methods can be used to analyse data which do not satisfy the requirements of conventional statistical methods. For example, it may be that a researcher would like to analyse the results of an experiment, using continuous attributes, by means of a multivariate analysis of variance but this is impossible because the system is over-defined. A solution might be to analyse the data by a principal component analysis followed by a set of univariate Anova's of the resulting components. With mixed data types principal co-ordinate analysis could be used and the resulting Gower vectors could be submitted to Anova. In the latter case, the problem is that the distribution of the values on the vectors is unknown so that significance tests would have to be interpreted cautiously. However, some sort of analysis would be possible on otherwise intractable data. Sequential results, such as weight gains over a number of time intervals, could be treated similarly. The main effects tend to separate on different vectors. A similar effect may be found in experiments in which soil heterogeneity exists (Williams 1976 pp 133–134).

It is often believed that the use of numerical methods makes a study more objective and thus more respectable scientifically. In fact, numerical methods are objective only in that they treat all data equally and eliminate personal bias during the execution of the methods. In environmental data analysis the subjective element enters at almost every stage, in the nature of the data to be collected, in the choice of numerical methods, and often in the use of the computer output, for example in choosing how many dimensions of an ordination to accept or in setting up clustering parameters. Any data analysis is likely to begin with some sort of assumption, even if it is only the simple one that there is some pattern to be found. The course of the analysis depends on the nature of the assumptions and on decisions made by the researcher as to what information is useful and what is not. As such decisions are likely to be made on the basis of the initial assumptions, the latter must be examined constantly and their validity checked. Otherwise, there is a danger that the answer expected by the researcher may be built into the results inadvertently.

It is a regrettable consequence of the availability of digital computers for 'hands-on' use by researchers that opportunities exist for data to be analysed by methods for which they are completely unsuitable. Normally, a computer program will check for certain major computational problems and will produce a failure message if they are found, for example a zero determinant in a matrix inversion program. If such computational problems are not found a program will produce an output, even if it is nonsense. Because there is in many minds a tendency to attribute to a computer the ability to produce 'objective truth', much time can be wasted on the interpretation of garbage output. The human mind is ingenious, and can usually find a plausible explanation for a set of results. If the results are subsequently shown to be erroneous, an equally plausible explanation may be found for a completely different set of results.

Here, the aim is to offer some broad guidelines which, it is hoped, will set the researcher thinking carefully about the properties of the different methods in relation to his/her own requirements. In order to avoid wasting time and possibly producing erroneous conclusions, some understanding of the potentials and problems of the methods is necessary. Therefore, the researcher should make some attempt to understand the properties of a numerical method before using it, and should not use, uncritically, any method which happens to be available.

The first important requirement is a clear statement of the objectives of the research. This should be done before the research is carried out, and in sufficient detail to enable a logical choice of objects and attributes to be made. Obviously, the data matrix must contain sufficient information adequately to portray the phenomenon of interest. Where the problem is clearly defined there is a need for problem-specific information. However, it is often the case that the researcher does not know which attributes are important. Although he/she may have reason to believe that certain attributes may be relevant, it may not be known if that selection is adequate.

Some researchers take the view that more is better, and include all possible attributes in the hope that something will emerge from the data analysis. This not only makes unnecessary work but can also confuse the analysis (see Section 3.2). Consider the cluster analysis of data obtained from map grid squares. It might seem like a good idea to include some measure of location for each square, such as the eastings and northings of the British National Grid Reference. However, would it be logical for two squares with otherwise similar properties to be assigned to different clusters because they occur in different locations? The presence of irrelevant or non-diagnostic attributes seriously reduces the effectiveness of clustering methods which use minimum-variance criteria or single-linkage procedures. This is compounded by data standardization and transformation (Wishart 1969b).

Johnson (1982) found that in cluster analysis of data for three *Avena* species, selectively reducing the number of attributes on the basis of their *F* values improved the effectiveness of the analysis and reduced the number of mixed groups. Orloci (1978) gave a useful program called RANK, which ranks attributes according to the independent share that they can possibly absorb of the total sum of squares.

The selection of objects is also important. Inadequate or biased sampling can result in artificial discontinuities in the distribution of the points representing objects in the multidimensional attribute space (van Groenewoud 1983). In Q-mode analysis, the number of attributes must be sufficient to provide stability in the similarities or dissimilarities used. In R-mode analysis there must be sufficient objects to bring about stability in the variances and covariances or the derived correlations.

At this stage, thinking about the methods by which the data will eventually be analysed will ensure that the data are collected in a way which is suitable for the preferred analysis, or alternatively that a method of analysis is available for the preferred method of recording the data. If the nature of the problem is such as to make the use of one particular method logical, the data need only satisfy the requirements of that method. On the other hand, if the problem is an exploratory one where the data need to be submitted to a variety of techniques, it should be borne in mind that some types of data restrict the range of methods which can be used.

Some properties are logically recorded as continuous attributes, so the data analysis method should be suitable for handling them, possibly mixed with other types. The widest range of methods available is for data matrices which are entirely of continuous attributes. Where there is a logical choice of recording attributes in different ways, the advantages and disadvantages of the different ways can be weighted against the advantages and disadvantages of the different methods for handling the data, as some methods are suitable for one type of data, other methods are better for another type. The conversion of continuous attributes to binary or multistate should be avoided as it wastes information and can introduce distortion. These aspects are discussed in Chapter 3.

Sparse data matrices can present problems. Covariances and correlation coefficients are biased when calculated from samples with many double-zeros. Also, the relationships between pairs of samples with many double-zeros may be distorted when Pythagorean distance is computed (e.g. see Legendre and Legendre 1983 p 195). Such data will give an unsatisfactory ordination with principal component analysis, but useful ordinations could be obtained by principal co-ordinate analysis, using a similarity coefficient that excludes double-zeros, and by correspondence analysis as the chi-square distance excludes double zeros (see Legendre and Legendre 1983 p188–189).

Orientation data, such as compass bearings, present considerable problems and are probably better omitted from a multivariate data matrix. Methods for handling such data are given in Batschelet (1981).

A comment which has been made by more than one statistician is that researchers tend to go directly to advanced numerical methods without an adequate preliminary exploration of the data. Simple bivariate plots show if attributes are linearly related or not. A histogram can be useful in illustrating the shape of a distribution and can show if there is multimodality. Boxplots of attributes show if there is skewness and also if there are outliers, and enable the effects of different transformations to be examined (McNeil 1977). In univariate statistics, it has long been common practice to plot a sample of measurements on normal probability paper before embarking on a formal statistical analysis. Marked skewness is shown up by simple curvature of the plot, and the presence of kurtosis or of outlying values may also be indicated. Healey (1968) proposed a multivariate version of this procedure, in which a distance is calculated for each point

$$\mathbf{D}^2 = (\mathbf{x} - \bar{\mathbf{x}})' \mathbf{S}^{-1} (\mathbf{x} - \bar{\mathbf{x}})$$

which is the equation of an ellipsoid in the p-dimensional attribute space (see e.g. Morrison 1967). These distances can be plotted on normal probability paper or they can be presented as a histogram.

Banfield and Gower (1980) showed how the row sums of a binary data matrix can help with the interpretation of an ordination of the data, can reveal discrepancies and screen clustering possibilities, and also permit the recovery of approximations to all the a, b, c, d, values.

One of the first steps in pattern analysis is deciding how to organize the data. The methods described in the preceding chapters are based on a data matrix of N objects (represented by the rows) and p attributes (represented by the columns). The analysis of the p attributes together, rather than in separate univariate tests, uses interrelationships among them, increasing the power of the analysis and making the conclusions of wider relevance. However, there can be a third dimension, the occasion on which the observations took place. This gives a data cube which can be sliced in various ways to provide two-dimensional data matrices (see Jöreskog *et al.* 1976).

Multivariate methods do not generally associate different attributes with different *a priori* importance values, so that all attributes are normally treated in the same way. It is possible for different attributes to be given different weightings by the user, but this is in general undesirable (see Section 3.3). Sometimes the attributes are such that several might reflect some single underlying feature; there is then an implicit weighting of that feature. An ordination will normally reduce the dimensionality and provide scores which can be used in place of the original attributes in subsequent analyses, removing the effects of such implicit weighting.

Choosing an ordination method

Apart from the particular preoccupations of phytosociologists, the choice of an ordination method is fairly simple, and is constrained largely by the

Table 16.1 Types of data suitable for the ordination methods described

Ordination method	Type of data
R-mode	
Principal component analysis	Continuous attributes on interval or ratio scales, some meristic data. If all attributes are binary then use SSPS matrix, not covariance or correlation matrix
Factor analysis	Continuous attributes on interval or ratio scales, some meristic data
Q-mode	
Principal co-ordinate analysis	Continuous, binary, multistate, or mixed
Q-mode factor analysis	Proportions or percentages
Correspondence analysis	Contingency tables, data with additive properties such as species abundances, binary attributes

nature of the data. If there is a good reason why a particular method should be used, this should be taken into account at the planning stage when the choice of attributes is made. Otherwise, the use of an ordination method may be forced on to the user by the nature of the data. The types of data required by the ordination methods described in detail in this book are given in Table 16.1. The individual methods have advantages and disadvantages which are discussed in the corresponding chapters.

The choice between R-mode and Q-mode methods depends primarily on whether the user wishes to base the analysis on relationships among attributes or objects. If an R-mode analysis is required, the choice between principal component and factor analysis is generally made in favour of principal component analysis. This is because the method is robust, simple to understand and use, and is reasonably easy to interpret. Factor analysis is based on certain ideas about the distribution of the variance among the objects and attributes (see Chapter 4), and is more difficult to apply and interpret.

With regard to Q-mode methods, if the aim is to obtain a Euclidean representation of non-Euclidean data, principal co-ordinate analysis should be used. Q-mode factor analysis enables certain hypotheses concerning the possible components of mixtures to be tested. Correspondence analysis allows the graphical representation of both objects and attributes on the same axes.

Choosing a cluster analysis method

Greig-Smith (1983) discussed clustering methods with regard to the special problems and objectives of vegetation analysis. The following remarks deal with more general problems.

There are many possible uses of cluster analysis, but the five main ones are:

(i) Hypothesis generation. Cluster analysis can be used to generate a typology and then models can be constructed to explain the typology. This assumes the existence of a 'real' typology. It is necessary to bear in mind that the data used to obtain a clustering cannot be used to test any hypothesis generated from it.

(ii) Hypothesis testing. If the hypothesis allows a prediction to be made about the pattern of dissimilarities among objects, then the prediction can be compared with the results of a cluster analysis. This is not a statistical test, but it can be regarded as supporting evidence.

(iii) Data reduction. If N objects can be clustered into $g < N$ groups, a more concise and comprehensible description of the observations is possible. That is, we seek simplification with minimal information loss. There are no assumptions about the 'reality' of the groups.

(iv) Prediction. Clusters may consist of objects which have properties in common other than those used to produce the clusters, so that the clusters have some predictive value.

(v) Stratification for future sampling. Stratification yields more efficient, and therefore more precise, estimators where the attributes being investigated are homogeneous within the strata but are heterogeneous overall. The more the within-strata variance is reduced the more efficient is the stratification. Such a stratification can be produced by minimum variance clustering (Golder and Yeomans 1973).

In selecting a cluster analysis method, the researcher must differentiate between searching for distinct clusters which may exist (classification) or imposing clusters for practical purposes when no distinct clusters exist (dissection). Classification may be a technique for generating hypotheses but dissection is not, because the data have been forced into a strait-jacket which restricts the domain of possible hypotheses and suggests that some will be generated by the dissection rather than by the data (Cormack 1971). In broad terms, we can associate hypothesis generation and testing with classification methods, while data reduction may be served equally well by classification or dissection, depending on the user's objectives. Stratification requires a dissection method with minimum variance properties, and such a method is likely to be useful for prediction unless the data are all binary, when Gower's (1974) maximal predictive classification should be useful.

In classification, hierarchical methods are appropriate if the main interest is in relationships among objects and clusters, as these methods optimize the dendrogram structure at the expense of the cluster compositions. For exploratory work the method chosen should not tend to impose a structure on the data, that is, it should be broadly space-conserving. One of the arithmetic average methods would be a logical choice, normally the unweighted method (UPGMA) which gives all objects equal weight and produces less distortion than the weighted method when dendrograms are compared with original similarity matrices. However, there are occasions when a weighted method (WPGMA) may be preferred (Sneath and Sokal 1973 p228). It is especially suitable in situations where the groups are represented by disparate sample sizes and it is preferable to give equal weights to the groups. Gower (1967a) suggested that this method is, perhaps, the best general-purpose method. An alternative to UPGMA is flexible clustering with $\beta = -0.25$.

However, the arithmetic average methods have minimum variance properties and may divide certain types of cluster in an unacceptable manner (see Chapter 13). Single-linkage cluster analysis will find clusters having a variety of shapes but it needs to be used thoughtfully and with some care. It is more likely to show up a lack of real structure than are the other methods.

In exploratory analysis, if conceptually different clustering methods tend to produce the same clusters there is good reason to suppose that some real structure has been found. If different methods produce different results, the latter must be interpreted with reference to the properties of the methods and the user's objectives. The results of any analysis need to be evaluated in relation to known, or hypothesized, facts and relationships. Researchers using exploratory methods of pattern analysis do not always proceed to hypothesis testing. Examples where this has been done are in Gittins (1979) and Goldsmith (1973a,b).

Many hierarchical methods can be used with a variety of dissimilarity or similarity measures. If the attributes are all binary (0,1), a similarity or association coefficient or Pythagorean distance will be appropriate. Sneath and Sokal (1973 p157) recommended that binary attributes should not be standardized. Continuous attributes (interval and ratio scale data) and some meristic attributes (Section 3.1) suggest the use of a distance or dissimilarity measure. Multistate data can be used to calculate similarity coefficients, as in Gower's general similarity coefficient. Mixed data types are most easily dealt with using Gower's general similarity coefficient. For qualitative attributes the user must decide if double-zero matches are to count or not. Ward's error sum of squares method is strictly Euclidean.

Non-hierarchical methods are preferable if the aim is to optimize the cluster composition at a given level, as in dissection. k-means clustering (Hartigan and Wong 1979) produces clusters with minimum variances, a property which is useful for many practical purposes and essential for some. A criterion which helps the user to decide on the optimum number of clusters can be computed. The method is strictly Euclidean.

Other clustering methods may be preferred for specific objectives, but the researcher will need to study the chapters on cluster analysis methods in order to understand their properties, advantages and disadvantages.

228

If the data consist of profiles one might consider using a profile-based method. Hawkins *et al.* (1982) discussed the options for normal profile data, or data which may be transformed readily to a form in which a normal mixture would fit reasonably well. With profiles of non-normal interval data, some ingenuity may be necessary to define a suitable dissimilarity or similarity coefficient. Optimization techniques are usually practicable only for small data sets, and the methods make some assumptions about the cluster shapes. Methods based on fitting mixtures of normal distributions run into many practical problems. Methods such as mode analysis rely on values of parameters which must be chosen by the user.

A decision which needs to be made is whether or not to perform a cluster analysis on the original attributes. In general, it can be advantageous to perform the cluster analysis on ordination scores, which usually allow the dimensionality to be reduced. With continuous attributes there is much to be said for ordination by principal component analysis, which removes the implicit weighting caused by redundant attributes and allows much of the 'noise' to be discarded. The component values can then be used in cluster analysis. With binary, multistate, or mixed data there is a choice of: (a) performing a principal co-ordinate analysis on a similarity matrix or a derived distance matrix, and a cluster analysis on the same matrix; and (b) using a principal co-ordinate analysis to obtain a Euclidean representation of non-Euclidean data, and performing a cluster analysis on the scores on the Gower vectors. The user's objectives will decide which course is to be taken.

Cluster analysis methods are essentially descriptive, the results should not be accepted blindly but should lead to a proper re-examination of the data. It is advantageous to have an ordination to provide a graphical display of the clusters. This enables unrealistic groupings to be shown up. An alternative form of graphical representation can be obtained by computing an intercluster distance matrix followed by a principal co-ordinate ordination. Andrews plots provide a useful graphical representation of the relationships of points in clusters (see Everitt 1978).

Provided that the requirement of homogeneous covariance matrices is satisfied, discriminant function analysis may be used after cluster analysis as an informal indicator of which attributes have contributed most to cluster formation, but individual attributes cannot be tested using the usual t or F tests (Everitt 1979). Stepwise discriminant analysis provides useful information such as the order of importance of attributes as discriminators and a matrix of F-statistics to test for significant differences between groups at each step (e.g. see Denton and del Moral 1976). Canonical variate analysis can be useful in displaying the groups of objects after a cluster analysis. However, the covariance matrices of the clusters are unlikely to be homogeneous if the clusters contain widely-different numbers of objects. If the application of discriminant function or canonical variate analysis and of tests of significance is an essential part of the objectives, then a clustering method with group-size dependence, such as flexible clustering, is required (Clifford and Stephenson 1975 p137). If there are problems with non-normality of distributions, a non-parametric test may be needed, such as the Kruskal-Wallis test, to test that the k clusters are, in fact, different. Such a test may be able to accommodate clusters with different numbers of objects.

A classification, in the broad sense, may be expected to predict attributes not included in its construction. Such attributes may be deliberately omitted from the construction so that they may be used as a test (e.g. Fisher *et al.* 1967; Watson *et al.* 1967; Clifford *et al.* 1969; Edye *et al.* 1970).

Additional information on clusters suggested by single-linkage cluster analysis can easily be obtained in supplementary analyses. For example, it is useful to calculate inter- and intra-cluster average distances to indicate the density of linkage. A table of the k nearest neighbours of each point ($k = 5$ is a useful number) reveals possible alternative branches in the event of ties and is useful in interpretation.

Batista (1988) described a technique, called discriminant co-ordinates analysis, for analysing the linear relationships between a number of new variates (e.g. environmental or functional attributes) and attributes (e.g. vegetation) which have been summarized in the form of a classification in the general sense of that term. The technique, which Batista illustrated with a case study, displays the principal differences among classes in relation to the new variates, and is equivalent to a special case of principal component analysis.

Many researchers like to see examples of data analysis in order to appreciate the possibilities of the methods. Sneath and Sokal (1973 Chap. 11) gave several examples of the application of pattern analysis methods in ecology, biogeography and earth sciences, which serve to illustrate the scope of the methods (see also Williams 1976). Jöreskog *et al.* (1976) gave some geological examples. It must be emphasized that a heavy interpretative load is placed on the user. Ecological data carry considerable random variation, and ecological problems are usually complex. As Orloci (1978) remarked, they cannot be reasoned out as neatly as the orbital paths of comets or the trajectory of a space module. The only real criterion is usefulness, which can only be tested empirically.

REFERENCES

Adamson, J.K. (1984). Land classification as a basis for stratified soil survey: a study in Cumbria. *Appl. Geog.*, **4**, 309–320.

Aitchison, J.W. (1978). Classification and mixed-mode data: An appreciation of Gower's general coefficient of similarity. *Cambria*, **5**, 145–155.

Allen, T.F.H., Sadowsky, D.A. and Woodhead, N. (1984). Data transformation as a scaling operation in ordination of plankton. *Vegetatio*, **56**, 147–160.

Allen, T.F.H. and Shugart, H.H. (1983). Ordination of simulated complex forest succession: A new test of ordination methods. *Vegetatio*, **51**, 141–155.

Anderberg, M.R. (1973). Cluster analysis for applications. New York; London: Academic Press, 359 pp.

Anderson, A.J.B (1971). Ordination methods in ecology. *J. Ecol.*, **59**, 713–726.

Anderson, T.W. (1958). An introduction to multivariate statistical analysis. New York; London: Wiley, 374 pp.

Anderson, T.W. (1963). Asymptotic theory for principal component analysis. *Ann. Math. Statist.*, **34**, 122–128.

Anderson, T.W. and Rubin, H. (1956). Statistical inference in factor analysis. In: Proc. Symp. Math. Statist. and Prob. Vol. 5, edited by J. Neyman, 111–150. Berkely: Univ. California Press.

Andrews, D.F., Gnanadesikan, R. and Warner, J.L. (1971). Transformations of multivariate data. *Biometrics*, **27**, 825–840.

Armstrong, J.S. (1967). Derivation of theory by means of factor analysis, or Tom Swift and his electric factor analysis machine. *Am. Statistn.*, Dec., 17–21.

Austin, M.P. (1968). An ordination study of a chalk grassland community. *J. Ecol.*, **56**, 739–757.

Austin, M.P. (1976). On non-linear response models in ordination. *Vegetatio*, **33**, 33–41.

Austin, M.P. (1980). Searching for a model for use in vegetation analysis. *Vegetatio*, **42**, 11–21.

Austin, M.P. (1985). Continuum concept, ordination methods, and niche theory. *Annu. Rev. Ecol. Syst.*, **16**, 39–61.

Austin, M.P. and Noy-Meir, I. (1971). The problem of non-linearity in ordination: Experiments with two-gradient models. *J. Ecol.*, **59**, 763–773.

Austin, M.P. and Orloci, L. (1966). Geometric models in ecology II. An evaluation of some ordination techniques. *J. Ecol.*, **54**, 217–227.

Baker, F.B. (1962). Information retrieval based on latent class analysis. *J. Ass. Comput. Mach.*, **9**, 512–521.

Baker, F.B. and Hubert, L.J. (1975). Measuring the power of hierarchical cluster analysis. *J. Am. Statist. Ass.*, **70**, 31–38.

Banfield, C.F. (1976). Algorithm AS102. Ultrametric distances for a single-linkage dendrogram. *Appl. Statist.*, **25**, 313–315.

Banfield, C.F. (1978). Singular value decomposition in multivariate analysis. In: Numerical software, needs and availability, edited by D. Jacobs, 137–149. London; New York: Academic Press.

Banfield, C.F. and Gower, J.C. (1980). A note on the graphical representation of multivariate binary data. *Appl. Statist.*, **29**, 238–245.

Barkham, J.P. (1968). The ecology of the ground flora of some Cotswold beechwoods. Ph.D. Thesis, University of Birmingham.

Barkham, J.P. and Norris, J.M. (1970). Multivariate procedures in an investigation of vegetation and soil relations of two beech woodlands, Cotswold Hills, England. *Ecology*, **51**, 630–639.

Bartlett, M.S. (1950). Tests of significance in factor analysis. *Br. J. Psych. (Statist. Sect.)*, **3**, 77–85.

Bassett, P.A. (1978). The vegetation of a Camargue pasture. *J. Ecol.*, **66**, 803–827.

Batista, W.P. (1988). Relating new information to a previous vegetation classification: a case of discriminant co-ordinates analysis. *Vegetatio*, **75**, 153–158.

Batschelet, E. (1981). Circular statistics in biology. London; New York: Academic Press, 371 pp.

Baum, B.R. (1978). Taxonomy of the tribe Triticeae (Poaceae) using various numerical techniques. II. Classification. *Can. J. Bot.*, **56**, 27–56.

Bauzon, D., Ponge, J.-F. and Dommergues, Y. (1974). Varitions saisonnierès des caractéristiques chimiques et biologiques des sols forestiers, interprétées par l'analyse factorielle des correspondances. *Rev. Ecol. Biol. Sol.*, **11**, 283–301.

Beale, E.M.L. (1969). Euclidean cluster analysis. 37th Session of the Int. Statist. Inst., 99–101.

Beals, E.W. (1973). Ordination: Mathematical elegance and ecological naivete. *J. Ecol.*, **61**, 23–35.

Beals, E.W. (1984). Bray-Curtis ordination: An effective strategy for analysis of multivariate ecological data. *Adv. Ecol. Res.*, **14**, 1–55.

Beeston, G. and Dale, M.B. (1975). Multiple predictive analysis: A management tool. *Proc. Ecol. Soc. Aust.*, **9**, 172–181.

Bell, C.R., Holder-Franklin, M.A. and Franklin, M. (1982). Seasonal fluctuations in river bacteria as measured by multivariate statistical analysis of continuous cultures. *Can. J. Microbiol.*, **28**, 959–975.

Benzécri, J.-P. (1969). Statistical analysis as a tool to make patterns emerge from data. In: Methodologies of pattern recognition, edited by S. Watanabe, 35–60. New York: Academic Press.

Benzécri, J.-P. (1973) L'Analyse des données. Vol. 2. L'Analyse des correspondances. Paris: Dunod, 619 pp.

Betters, D.R. and Rubingh, J.L. (1978). Suitability analysis and wildland classification: an approach. *J. Environ. Manage.*, **7**, 59–72.

Bezdeck, J.C. (1974). Numerical taxonomy with fuzzy sets. *J. Math. Biol.*, **1**, 57–71.

Birks, H.J.B. (1974). Numerical zonations of Flandrian pollen data. *New Phytol.*, **73**, 351–358.

Blackith, R.E. and Reyment, R.A. (1971). Multivariate morphometrics. London; New York: Academic Press, 412 pp.

Bonin, G. and Roux, M. (1978). Utilisation de l'analyse factorielle des correspondances dans l'étude phyto-écologique de quelques pelouses de l'Appenin lucano-calabrias. *Oecol. Plant.*, **13**, 121–138.

Bonnet, L., Cassagnau, P. and Travé, J. (1975). L'écologie des arthropodes muscioles à la lumiere de l'analyse des correspondances: Collemboles et oribates du Sidobre (Tarn, France). *Oecologia*, **21**, 359–373.

Booth, T.H. (1978). Numerical classification techniques applied to forest tree distribution data. I. A comparison of methods. *Aust. J. Ecol.*, **3**, 297–306.

Boratynski, K. and Davies, R.G. (1971). The taxonomic value of male Coccoidea (Homoptera) with an evaluation of some numerical techniques. *Bot. J. Linn. Soc.*, **3**, 57–102.

Bottomley, J. (1971). Some statistical problems arising from the use of the information statistic in numerical classification. *J. Ecol.*, **59**, 339–342.

Bouxin, G. (1976). Ordination and classification in the upland Rugeye forest (Rwanda, Central Africa). *Vegetatio*, **32**, 97–115.

Bowdler, H., Martin, R.S., Reinsch, C. and Wilkinson, J.H. (1971). The QR and QL algorithms for symmetric matrices. In: Handbook for automatic computation Vol. II Linear algebra, edited by J.H. Wilkinson and C. Reinsch, 227–240. Berlin; Heidelberg; New York: Springer-Verlag.

Boyce, A.J. (1969). Mapping diversity: A comparative study of some numerical methods. In: Numerical taxonomy, edited by A.J. Cole, 1–31. London; New York: Academic Press.

Bray, J.R. and Curtis, J.T. (1957). An ordination of the upland forest communities of southern Wisconsin. *Ecol. Monogr.*, **27**, 325–349.

Brown, M.J., Ratkowsky, D.A. and Minchin, P.R. (1984). A comparison of detrended correspondence analysis and principal co-ordinates analysis using four sets of Tasmanian vegetation data. *Aust. J. Ecol.*, **9**, 273–279.

Bunce, R.G.H., Morrell, S.K. and Stel, H.E. (1975). The application of multivariate analysis to regional survey. *J. Environ. Manage.*, **3**, 151–165.

Burns, D. (1983). Graph theory. In: Encyclopedia of statistical sciences Vol. 3, edited by S. Kotz and N.L. Johnson, 517–522. New York: Wiley.

Burr, E.J. (1968). Cluster sorting with mixed character types. I. Standardisation of character values. *Aust. Comput. J.*, **1**, 97–99.

Burr, E.J. (1970). Cluster sorting with mixed character types. II. Fusion strategies. *Aust. Comput. J.*, **2**, 98–103.

Cailliez, F. and Pagés, J.P. (1976). Introduction a l'analyse des données. Paris: Societe de Mathematiques appliquées et de Sciences humaines, 616 pp.

Cameron, E.M. (1968). A geochemical profile of the Swan Hills Reef. *Can. J. Earth Sci.*, **5**, 287–309.

Carleton, T.J. (1980). Non-centered component analysis of vegetation data: a comparison of orthogonal and oblique rotation. *Vegetatio*, **42**, 59–66.

Carroll, J.B. (1958). Oblimin rotation solution in factor analysis. Harvard Univ: computing program for the IBM 704.

Carroll, J.B. (1961). The nature of the data, or how to choose a correlation coefficient. *Psychometrika*, **26**, 347–372.

Carroll, J.D. (1972). Individual differences and multidimensional scaling. In: Multidimensional scaling, theory and applications in the behavioural sciences, Vol. I Theory, edited by R.N. Shepherd, A.K. Romney and S.B. Nerlove, 105–155. New York: Seminar.

Carroll, J.D. and Chang, J.J. (1970). Analysis of individual differences in multidimensional scaling via an N-way generalization of 'Eckart-Young' decomposition. *Psychometrika*, **35**, 283–320.

Cassie, R.M. (1963). Multivariate analysis in the interpretation of numerical plankton data. *N.Z.J. Sci.*, **6**, 36–59.

Cassie, R.M. (1969). Multivariate analysis in ecology. *Proc. N.Z. Ecol. Soc.*, **16**, 53–57.

Cassie, R.M. and Michael, A.D. (1968). Fauna and sediments of an intertidal mud flat: a multivariate analysis. *J. Exp. Mar. Biol. Ecol.*, **2**, 1–23.

Cattell, R.B. (1952). Factor analysis. New York: Harper, 462 pp.

Cattell, R.B. (1965). Factor analysis: An introduction to essentials I. The purpose and underlying models. *Biometrics*, **21**, 190–215.

Cattell, R.B. (1966). Handbook of multivariate experimental psychology. Chicago: Rand McNally.

Cattell, R.B. and Khanna, D.K. (1977). Principles and procedures for unique rotation in factor analysis. In: Mathematical methods for digital computers Vol. III, edited by K. Enslein, A. Ralston and H.S. Wilf, 166–202. New York; London: Wiley.

Cattell, R.B. and Muerle, J.L. (1960). The 'maxplane' program for factor rotation to oblique simple structure. *Educ. Psychol. Measmt.*, **20**, 569–590.

Chambers, J.M. (1977). Computational methods for data analysis. New York; London: Wiley, 268 pp.

Chang, W.C. (1983). On using principal components before separating a mixture of two multivariate normal distributions. *Appl. Statist.*, **32**, 267–275.

Chardy, P., Glemarec, M. and Laurec, A. (1976). Application of inertia methods to benthic marine ecology: practical implications of the basic options. *Estuarine Coastal Mar. Sci.*, **4**, 179–205.

Chatfield, C. and Collins, A.J. (1980). Introduction to multivariate analysis. London; New York: Chapman and Hall, 246 pp.

Chayes, F. and Kruskal, W. (1966). An approximate statistical test for correlations between proportions. *J. Geol.*, **74**, 692–702.

Cheetham, A.H. and Hazel, J.E. (1969). Binary (presence-absence) similarity coefficients. *J. Palaeontol.*, **43**, 1130–1136.

Clifford, H.T. and Stephenson, W. (1975). An introduction to numerical classification. New

York; London: Academic Press, 229 pp.

Clifford, H.T. and Williams, W.T. (1980). Interrelationships among the Liliatae: A graph theory approach. *Aust. J. Bot.*, **28**, 261–268.

Clifford, H.T., Williams, W.T. and Lance, G.N. (1969). A further numerical contribution to the classification of Poaceae. *Aust. J. Bot.*, **17**, 119–131.

Cochran, W.G. (1963). Sampling techniques. New York; London: Wiley. 2nd edition, 413 pp.

Cohen, A., Gnanadesikan, R., Kettenring, J.R. and Landwehr, J.M. (1977). Methodological developments in some applications of clustering. In: Proc. Symp. on Applications of Statistics, edited by P.R. Krishnaiah, 141–162. Amsterdam: North-Holland.

Cole, A.J. and Wishart, D. (1970). An improved algorithm for the Jardine-Sibson method of generating overlapping clusters. *Comput. J.*, **13**, 156–163.

Comrey, A.L. (1973). A first course in factor analysis. New York: Academic Press.

Cooley, W.W. and Lohnes, P.R. (1971). Multivariate data analysis. New York; London: Wiley, 364 pp.

Cooper, A. (1984). Application of multivariate methods to a study of community composition and structure in an escarpment woodland in northeast Ireland. *Vegetatio*, **55**, 93–104.

Corbet, G.B., Cummins, J., Hedges, S.R. and Krzanowski, W. (1970). The taxonomic status of British water voles, genus Arvicola. *J. Zool.*, **161**, 301–316.

Cormack, R.M. (1971). A review of classification. *J. R. Statist. Soc. Ser. A.*, **134**, 321–367.

Coxon, A.P.M. (1982). The users guide to multidimensional scaling. London: Heinemann.

Craddock, J.M. (1973). Problems and prospects for eigenvector analysis in meteorology. *The Statistician*, **12**, 133–145.

Crawford, C.B. and Ferguson, G.A. (1970). A general rotation criterion and its use in orthogonal rotation. *Psychometrika*, **35**, 321–332.

Crawford, R.M.M. and Wishart, D. (1967). A rapid multivariate method for the detection and classification of groups of ecologically related species. *J. Ecol.*, **55**, 505–524.

Crawford, R.M.M. and Wishart, D. (1968). A rapid classification and ordination method and its application to vegetation mapping. *J. Ecol.*, **56**, 385–404.

Cronbach, L.J. (1951). Coefficient alpha and the internal structure of tests. *Psychometrika*, **16**, 297–334.

Cunningham, K.M. and Ogilvie, J.C. (1972). Evaluation of hierarchical grouping techniques: A preliminary study. *Comput. J.*, **15**, 209–213.

Cureton, E.E. and d'Agostino, R.B. (1983). Factor analysis, an applied approach. Hillsdale, N.J.: Erlbaum, 457 pp.

Dale, M.B. (1964). The application of multivariate methods to heterogeneous data. Ph.D. Thesis, Univ. of Southampton.

Dale, M.B. (1975). On objectives of methods of ordination. *Vegetatio*, **30**, 15–32.

Dale, M.B. (1977a). Graph theoretical analysis of the phytosociological structure of plant communities: the theoretical basis. *Vegetatio*, **34**, 137–154.

Dale, M.B. (1977b). Graph theoretical analysis of the phytosociological structure of plant communities: an application to mixed forest. *Vegetatio*, **35**, 35–46.

Daling, J.R. and Tamura, H. (1970). Use of orthogonal factors for selection of variables in a regression equation – an illustration. *Appl. Statist.*, **19**, 260–268.

Daniels, R.E. (1978). Floristic analyses of British mires and mire communities. *J. Ecol.*, **66**, 773–802.

Darton, R.A. (1980). Rotation in factor analysis. *The Statistician*, **29**, 167–194.

David, M., Campiglio, C. and Darling, R. (1974). Progresses in R- and Q-mode analysis: correspondence analysis and its application to the study of geological processes. *Can. J. Earth Sci.*, **11**, 131–146.

Davies, P.M. and Coxon, A.P.M. (eds.) (1982). Key texts in multidimensional scaling. London: Heinemann.

Davis, J.C. (1973). Statistics and data analysis in geology. New York; London: Wiley. 550 pp.

Day, N.E. (1969). Estimating the components of a mixture of normal distributions. *Biometrika*, **56**, 463–474.

Dean, W.E. and Gorham, E. (1976). Classification of Minnesota lakes by Q- and R-mode factor analysis of sediment mineralogy and geochemistry. In: Quantitative techniques for the analysis of sediments, edited by D.F. Merriam, 61–71. Oxford; New York: Pergamon Press.

Defays, D. (1977). An efficient algorithm for a complete link method. *Computer J.*, **20**, 364–366.

Delany, M.J. (1965). Application of factor analysis to the study of variation in the long-tailed fieldmouse (*Apodemus sylvaticus* (L.)) in north-west Scotland. *Proc. Linn. Soc. Lond.*, **176**, 103–111.

Denton, M.F. and del Moral, R. (1976). Comparison of multivariate analyses using taxonomic data of *Oxalis*. *Can. J. Bot.*, **54**, 1637–1646.

Dighton, J., Poskitt, J.M. and Howard, D.M. (1986). Changes in occurrence of basidiomycete fruit bodies during forest stand development: with specific reference to mycorrhizal species. *Trans. Br. Mycol. Soc.*, **87**, 163–171.

Duncan, T. and Estabrook, D.F. (1977). An operational method for evaluating classifications. *Syst. Bot.*, **1**, 373–382.

Eckart, C. and Young, G. (1936). The approximation of one matrix by another of lower rank. *Psychometrika*, **1**, 211–218.

Edwards, A.W.R. and Cavalli-Sforza, L.L. (1965). A method for cluster analysis. *Biometrics*, **21**, 362–375.

Edye, L.A., Williams, W.T. and Pritchard, A.J. (1970). A numerical analysis of variation pattern in Australian introductions of *Glycine wightii* (*G. javonica*). *Aust. J. Agric. Res.*, **21**, 57–70.

Elffers, H. (1980). On uninterpretability of factor analysis results. *Trans. Inst. Br. Geogr.*, **5**, 318–330.

Estabrook, G.F. (1966). A mathematical model in graph theory for biological classification. *J. Theor. Biol.*, **12**, 297–310.

Everett, J.E. (1983). Factor comparability as a means of determining the number of factors and their rotation. *Multivar. Behav. Res.*, **18**, 197–218.

Everitt, B.S. (1974). Cluster analysis. London: Heinemann, 122 pp. (Reprinted 1977).

Everitt, B.S. (1978). Graphical techniques for multivariate data. London: Heinemann, 117 pp.

Everitt, B.S. (1979). Unresolved problems in cluster analysis. *Biometrics*, **35**, 169–182.

Everitt, B.S. (1980). Cluster analysis. London: Heinemann, 2nd edition.

Faith, D.P. (1983). Asymmetric binary similarity measures. *Oecologia*, **57**, 287–290.

Faith, D.P., Minchin, P.R. and Belbin, L. (1987). Compositional dissimilarity as a robust measure of ecological distance. *Vegetatio*, **69**, 57–68.

Farris, J.S. (1969). On the cophenetic correlation coefficient. *Syst. Zool.*, **18**, 279–285.

Farris, J.S. (1973). On comparing the shapes of taxonomic trees. *Syst. Zool.*, **22**, 50–54.

Farris, J.S. (1977). On the phenetic approach to vertebrate classification. In: Major patterns in vertebrate evolution, edited by M.K. Hecht, P.C. Goody and B.M. Hecht. New York: Plenum Press.

Farris, J.S., Kluge, A.G. and Eckart, M.J. (1970). A numerical approach to phylogenetic systematics. *Syst. Zool.*, **19**, 172–189.

Fasham, M.J.R. (1977). A comparison of non-metric multidimensional scaling, principal components and reciprocal averaging for the ordination of simulated coenoclines and coenoplanes. *Ecology*, **58**, 551–561.

Feoli, E. (1976). Correlation between single ecological variables and vegetation by means of cluster analysis. *Not. Fitosoc.*, **12**, 77–82.

Feoli, E. and Gerdol, R. (1982). Evaluation of syntaxonomic schemes of aquatic plant communities by cluster analysis. *Vegetatio*, **49**, 21–27.

Fewster, P.H. and Orloci, L. (1983). On choosing a resemblance measure for non-linear predictive ordination. *Vegetatio*, **54**, 27–35.

Field, J.G. (1969). The use of the information statistic in the numerical classification of heterogeneous systems. *J. Ecol.*, **57**, 565–569.

Fisher, L. and van Ness, J.W. (1971). Admissible clustering procedures. *Biometrika*, **58**, 91–104.

Fisher, R.A. (1936). The use of multiple measures in taxonomic problems. *Ann. Eugen.*, **8**, 376–386.

Fisher, R.A. (1940). The precision of discriminant functions. *Ann. Eugen.*, **10**, 422–429.

Fisher, R.G., Williams, W.T. and Lance, G.N. (1967). An application of techniques of numerical taxonomy to company information. *Econ. Rec.*, **43**, 566–587.

Fisher, W.D. (1969). Clustering and aggregation in economics. Baltimore: Johns Hopkins Press.

Fitch, W.M. (1977). On the problem of discovering the most parsimonious tree. *Am. Nat.*, **111**, 223–257.

Forgey, E.W. (1964). Evaluation of several methods for detecting sample mixtures for different N-dimensional popuations. *Amer. Psychol. Assn. Meetings*, Los Angeles, California.

Forgey, E.W. (1965). Cluster analysis of multivariate data: efficiency versus interpretability of classifications. AAAS-Biometric Soc. Meetings (WNAR), Riverside, California.

Friedman, H.P. and Rubin, J. (1967). On some invariant criteria for grouping data. *J. Am. Statist. Ass.*, **62**, 1159–1178.

Gabriel, K.R. (1971). The biplot graphic display of matrices with application to principal component analysis. *Biometrika*, **58**, 453–467.

Gabriel, K.R. (1981). Biplot display of multivariate matrices for inspection of data and diagnosis. In: Interpreting multivariate data, edited by V. Barnett, 147–173. Chichester; New York: Wiley.

Gabriel, K.R. and Sokal, R.R. (1969). A new statistical approach to geographic variation analysis. *Syst. Zool.*, **18**, 259–278.

Gardner, G. (1978). A discussion of the techniques of cluster analysis with reference to an application in the field of social services. M.Sc. dissertation, University of Kent at Canterbury.

Gauch, H.G. and Wentworth, T.R. (1976). Canonical correlation analysis as an ordination technique. *Vegetatio*, **33**, 17–22.

Gauch, H.G. and Whittaker, R.H. (1972a). Ceonocline simulation. *Ecology*, **53**, 446–451.

Gauch, H.G. and Whittaker, R.H. (1972b). Comparison of ordination techniques. *Ecology*, **53**, 868–875.

Gauch, H.G. and Whittaker, R.H. (1981). Hierarchical classification of community data. *J. Ecol.*, **69**, 537–557.

Gauch, H.G., Whittaker, R.H. and Singer, S.B. (1981). A comparative study of nonmetric ordinations. *J. Ecol.*, **69**, 135–152.

Gauch, H.G., Whittaker, R.H. and Wentworth, T.R. (1977). A comparative study of reciprocal averaging and other ordination techniques. *J. Ecol.*, **65**, 157–174.

Gitman, I. and Levine, M.D. (1970). An algorithm for detecting unimodal fuzzy sets and its application as a clustering technique. *IEEE Trans. on Computers*, **C-19**, 583–593.

Gittins, R. (1979). Ecological applications of canonical analysis. In: Multivariate methods in ecological work, edited by L. Orloci, C.R. Rao and W.M. Stiteler, 309–535. Fairland, Maryland: Int. Co-operative Publishing House.

Gnanadesikan, R. (1977). Methods for statistical data analysis of multivariate observations. New York; London: Wiley, 310 pp.

Golden, M.S. (1979). Forest vegetation of the lower Alabama piedmont. *Ecology*, **60**, 770–782.

Golder, P.A. and Yeomans, K.A. (1973). The use of cluster analysis for stratification. *Appl. Statist.*, **22**, 213–219.

Goldsmith, F.B. (1973a). The vegetation of exposed sea cliffs at South Stack, Anglesey. I. The multivariate approach. *J. Ecol.*, **61**, 787–818.

Goldsmith, F.B. (1973b). The vegetation of exposed sea cliffs at South Stack, Anglesey. II. Experimental Studies. *J. Ecol.*, **61**, 819–830.

Golub, G.H. and Reinsch, C. (1971). Singular value decomposition and least squares solutions. In: Handbook for automatic computation Vol. II. Linear algebra, edited by J.H. Wilkinson and C. Reinsch, 134–151. Berlin; Heidelberg: Springer-Verlag.

Goodall, D.W. (1966a). A new similarity index based on probability. *Biometrics*, **22**, 882–907.

Goodall, D.W. (1966b). The nature of the mixed community. *Proc. Ecol. Soc. Aust.*, **1**, 84–96.

Gordon, A.D. (1980). Methods of constrained classification. In: Analyse de données et informatique, edited by R. Tomassone, 161–171. Le Chesnay: IRIA.

Gordon, A.D. (1981). Classification. London; New York: Chapman & Hall, 193 pp.

Gordon, A.D. (1982). Numerical methods in Quaternary palaeoecology. V. Simultaneous graphical representation of the levels and taxa in a pollen diagram. *Rev. Palaeobot. Palynol.*, **37**, 155–183.

Gordon, A.D. and Birks, H.J.B. (1972). Numerical methods in Quaternary palaeoecology. I. Zonation of pollen diagrams. *New Phytol.*, **71**, 961–979.

Gordon, A.D. and Birks, H.J.B. (1974). Numerical methods in Quaternary palaeoecology. II. Comparison of pollen diagrams. *New Phytol.*, **73**, 221–249.

Gould, S.J. (1967). Evolutionary patterns in pelycosaurian reptiles: a factor analytical study. *Evolution*, **21**, 385–401.

Gould, S.J. (1969). An evolutionary microcosm: Pleistocene and recent history of the Land Snail. P. (Poecilozonites) in Bermuda. *Bull. Mus. Comp. Zool. Harv.*, **138**, 407–532.

Gower, J.C. (1966). Some distance properties of latent root and vector methods used in multivariate analysis. *Biometrika*, **53**, 325–338.

Gower, J.C. (1967a). A comparison of some methods of cluster analysis. *Biometrics*, **23**, 623–637.

Gower, J.C. (1967b). Multivariate analysis and multidimensional geometry. *The Statistician*, **17**, 13–28.

Gower, J.C. (1968). Adding a point to vector diagrams in multivariate analysis. *Biometrika*, **55**, 582–585.

Gower, J.C. (1969). A survey of numerical methods useful in taxonomy. *Acarologia*, **11**, 357–375.

Gower, J.C. (1970). A note on Burnaby's character-weighted similarity coefficient. *Math. Geol.*, **2**, 39–45.

Gower, J.C. (1971a). A general coefficient of similarity and some of its properties. *Biometrics*, **27**, 857–871.

Gower, J.C. (1971b). Statistical methods of comparing different multivariate analyses of the same data. In: Mathematics in the archaeological and historical sciences, edited by F.R. Hodson, D.G. Kendall and P. Tautu, 138–149. Edinburgh: Edinburgh Univ. Press.

Gower, J.C. (1972). Measures of taxonomic distance and their analysis. In: The assessment of population affinities in man, edited by J.S. Weiner and J. Huizinga, 1–24. Oxford: Clarendon Press.

Gower, J.C. (1974). Maximal predictive classification. *Biometrics*, **30**, 643–654.

Gower, J.C. (1975a). Generalized Procrustes analysis. *Psychometrika*, **40**, 33–51.

Gower, J.C. (1975b). Goodness-of-fit criteria for classification models and other patterned structures. In: Proc. 8th Int. Biometric Conf. Constanza, 38–62. London: Freeman.

Gower, J.C. (1985). Measures of similarity, dissimilarity, and distance. In: Encyclopedia of statistical sciences Vol. 5, edited by N.L. Johnson and S. Kotz, 397–405. New York: Wiley.

Gower, J.C. and Ross, G.J.S. (1969). Minimum spanning trees and single linkage cluster analysis. *Appl. Statist.*, **18**, 54–64.

Green, B.F. (1976). On the factor score controversy. *Psychometrika*, **41**, 263–266.

Green, P.E. and Rao, V.R. (1972). Applied multidimensional scaling, a comparison of approaches and algorithms. New York: Holt, Rinehart & Winston.

Green, P.E. and Tull, D.S. (1975). Research for marketing decisions. Englewood Cliffs, N.J.: Prentice-Hall, 3rd edition.

Greenacre, M.J. (1978). Some objective methods of graphical display of a data matrix. Univ. South Africa, Dept. of Statistics & Operations Research Special Report, 241 pp.

Greenacre, M.J. (1981). Practical correspondence analysis. In: Interpreting multivariate data, edited by V. Barnett, 119–146. Chichester; New York: Wiley.

Greenacre, M.J. (1984). Theory and applications of correspondence analysis. London: Academic Press.

Greenacre, M.J. and Degos, L. (1977). Correspondence analysis of HLA gene frequency data from 124 population samples. *Am. J. Hum. Genet.*, **29**, 60–75.

Greenacre, M.J. and Underhill, L.G. (1982). Scaling a data matrix in a low-dimensional Euclidean space. In: Topics in applied multivariate analysis, edited by D.M. Hawkins, 183–268. Cambridge; London: Cambridge Univ. Press.

Greenacre, M.J. and Vrba, E.S. (1984). Graphical display and interpretation of antelope census data in African wildlife areas using correspondence analysis. *Ecology*, **65**, 984–997.

Greig-Smith, P. (1964). Quantitative plant ecology, London: Butterworth, 2nd edition, 256 pp.

Greig-Smith, P. (1983). Quantitative plant ecology. Oxford: Blackwell, 3rd edition, 359 pp.

Guertin, W.H. and Bailey, J.P. (1970). Introduction to modern factor analysis. Ann Arbor: Edwards.

Guttman, L. (1953). Image theory for the structure of quantitative variates. *Psychometrika*, **18**, 277–296.

Guttman, L. (1954). Some necessary conditions for common factor analysis. *Psychometrika,* **19**, 149–161.

Guttman, L. (1955). The determinacy of factor scores matrices with implications for five other basic problems of common factor theory. *Br. J. Statis. Psychol.,* **8**, 65–81.

Hair, J.F., Anderson, R.E., Tatham, R.L. *et al.* (1979). Multivariate data analysis with readings. Tulsa: Petroleum Publishing Co.

Hall, J.B. and Swaine, M.D. (1976). Classification and ecology of closed canopy forest in Ghana. *J. Ecol.,* **64**, 913–951.

Hansen, P. and Delattre, M. (1978). Complete-link cluster analysis by graph colouring. *J. Am. Statist. Assn.,* **73**, 397–403.

Harman, H.H. (1967). Modern factor analysis. Chicago: Chicago Univ. Press, 2nd edition, 474 pp.

Harman, H.H. (1977). Minres method of factor analysis. In: Mathematical methods for digital computers Vol. III, edited by K. Enslein, A. Ralston and H.S. Wilf, 154–165. New York; London: Wiley.

Harris, C.W. (1962). Some Rao-Guttman relationships. *Psychometrika,* **27**, 247–263.

Harris, C.W. (1967). On factors and factor scores. *Psychometrika,* **32**, 363–379.

Harris, R.J. (1975). A primer of multivariate statistics. New York; London: Academic Press. 332 p.

Hartigan, J.A. (1967). Representation of similarity matrices by trees. *J. Am. Statist. Ass.,* **62**, 1140–1158.

Hartigan, J.A. (1975). Clustering algorithms. New York; London: Wiley, 351 pp.

Hartigan, J.A. (1977). Distribution problems in clustering. In: Classification and clustering, edited by J. van Ryzin, 45–71. New York: Academic Press.

Hartigan, J.A. and Wong, M.A. (1979). Algorithm AS 136. A K-means clustering algorithm. *Appl. Statist.,* **28**, 100–108.

Hawkins, D.M. (1973). On the investigation of alternative regressions by principal component analysis. *Appl. Statist.,* **22**, 275–286.

Hawkins, D.M. and Fatti, L.P. (1984). Exploring multivariate data using the minor principal components. *The Statistician,* **33**, 325–338.

Hawkins, D.M. and Kass, G.V. (1982). Automatic interaction detection. In: Topics in applied multivariate analysis, edited by D.M. Hawkins, 269–302. Cambridge; London: Cambridge Univ. Press.

Hawkins, D.M., Muller, M.W. and ten Krooden, J.A. (1982). Cluster analysis. In: Topics in applied multivariate analysis, edited by D.M. Hawkins, 303–356. Cambridge; London: Cambridge University Press.

Hazel, J.E. (1970). Binary coefficients and clustering in biostratigraphy. *Bull. Geol. Soc. Am.,* **81**, 3237–3252.

Healy, M.J.R. (1968). Multivariate normal plotting. *Appl. Statist.,* **17**, 157–161.

Hendrickson, A.E. and White, P.O. (1964). PROMAX: a quick method for rotation to oblique simple structure. *Br. J. Statist. Psychol.,* **17**, 65–70.

Henrich, C.J. (1980). Floating-point arithmetic: Can it be trusted? *Mini-Micro Systs.,* **13**, 143–151.

Hill, M.O. (1973). Reciprocal averaging: an eigenvector method of ordination. *J. Ecol.,* **61**, 237–249.

Hill, M.O. (1974). Correspondence analysis: a neglected multivariate method. *Appl. Statist.,* **23**, 340–354.

Hill, M.O., Bunce, R.G.H. and Shaw, M.W. (1975). Indicator species analysis, a divisive polythetic method of classification, and its application to a survey of native pinewoods in Scotland. *J. Ecol.,* **63**, 597–613.

Hill, M.O. and Gauch, H.G. (1980). Detrended correspondence analysis: An improved technique. *Vegetatio,* **42**, 47–58.

Holder-Franklin, M.A. and Wuest, L.J. (1983). Factor analysis as an analytical method in microbiology. In: Mathematics in microbiology, edited by M. Bazin, 139–169. London: Academic Press.

Holgersson, M. and Jorner, U. (1978). Decomposition of a mixture into normal components: A review. *Int. J. Bio-Med. Comput.,* **9**, 367–392.

Holland, D.A. (1969). Component analysis: an aid to the interpretation of data. *Expl. Agric.,* **5**, 151–164.

Hope, K. (1968). Methods of multivariate analysis. London: Univ. of London Press, 288 pp.

Hopke, P.K. (1976). The application of multivariate analysis for interpretation of the chemical and physical analysis of lake sediments. *J. Environ. Sci. Health,* **A11** (6), 367–383.

Horst, P. (1965). Factor analysis of data matrices. New York: Holt, Rinehart and Wilson, 730 pp.

Howard, P.J.A., Hornung, M. and Howard, D.M. (1980). Recording soil profile descriptions for computer retrieval. Merlewood R&D Paper, No. 78, 47 pp.

Howard, P.J.A. and Howard, D.M. (1981). Multivariate analysis of map data: A case-study in classification and dissection. *J. Environ. Manage.,* **13**, 23–40.

Howard, P.J.A. and Howard, D.M. (1988). Classification and dissection of environmental data using qualitative and mixed data types. *J. Environ. Manage.,* **26**, 313–319.

Howarth, R.J. and Murray, J.W. (1969). The foraminiferida of Christchurch Harbour, England: A reappraisal using multivariate techniques. *J. Paleont.,* **43**, 660–675.

Hubalek, Z. (1982). Coefficients of association and similarity based on binary (presence-absence) data: An evaluation. *Biol. Rev.,* **57**, 669–689.

Hubert, L.J. (1974). Approximate evaluation techniques for the single-link and complete-link hierarchical clustering procedures. *J. Am. Statist. Assn.,* **69**, 698–704.

Hunt, B., Marin, J. and Stone, P.J. (1966). Experiments in induction. New York: Academic Press.

Ibanez, F. (1973). Méthode d'analyse spatio-temporelle du processus d'échantillonage en planctologie, son influence dans l'interpretation des données par l'analyse en composantes principales. *Ann. Inst. Oceanogr. (Paris),* **49**, 83–111.

Imbrie, J. (1963). Factor and vector analysis programs for analyzing geologic data. *Office Naval Res. Geogr. Branch.,* Tech. Rep. No. 6, 83 pp.

Imbrie, J. and van Andel, T.H. (1964). Vector analysis of heavy mineral data. *Bull. Geol. Soc. Am.,* **75**, 1131–1155.

Imbrie, J. and Purdy, E. (1962). Classification of modern Bahamian carbonate sediments. *Am. Ass. Petrol. Geol. Mem.,* No. 7, 253–272.

Ingwersen, F. (1983). Numerical analysis of the timbered vegetation in Tidbinbilla nature reserve, A.C.T., Australia. *Vegetatio,* **51**, 157–179.

Ivimey-Cook, R.B. and Proctor, M.C.F. (1967). Factor analysis of data from an east Devon heath: A comparison of principal component and rotation solutions. *J. Ecol.,* **55**, 405–413.

Ivimey-Cook, R., Proctor, M.C.F. and Wigston, D.L. (1969). On the problem of 'R/Q' terminology in multivariate analysis of biological data. *J. Ecol.,* **57**, 673–676.

Jambu, M. and Lebeaux, M.O. (1983). Cluster analysis and data analysis. Amsterdam; Oxford: North-Holland, 898 pp.

James, A.T. (1977). Tests for a prescribed subspace of principal components. In: Multivariate analysis IV, edited by P.R. Krishnaiah, 73–77. Amsterdam: North-Holland.

James, G. and James, R.C. (1968). Mathematics dictionary. Princeton, New Jersey: Van Nostrand, 3rd edition 517 pp.

Jancey, R.C. (1974). Algorithm for detection of discontinuities in data sets. *Vegetatio,* **29**, 131–133.

Jardine, N. and Sibson, R. (1968). The construction of hierarchic and non-hierarchic classifications. *Comput. J.,* **11**, 117–184.

Jardine, N. and Sibson, R. (1971). Mathematical taxonomy. London; New York: Wiley, 286 pp.

Jeffers, J.N.R. (1965). Principal component analysis in taxonomic research. For Comm. Statist. Section Paper No. 83, 21 pp.

John, D.M., Lieberman, D. and Lieberman, M. (1977). A quantitative study of the structure and dynamics of benthic subtidal algal vegetation in Ghana (Tropical West Africa). *J. Ecol.,* **65**, 497–521.

Johnson, P.O. (1949). Statistical methods in research. Englewood Cliffs, N.J.: Prentice-Hall.

Johnson, R.M. (1963). On a theorem stated by Eckart and Young. *Psychometrika,* **28**, 259–263.

Johnson, R.W. (1982). Effect of weighting and size of the attribute set in numerical classification. *Aust. J. Bot.,* **30**, 161–174.

Johnson, R.W. and Goodall, D.W. (1979). A maximum-likelihood approach to non-linear ordination. *Vegetatio,* **41**, 133–142.

Johnson, S.C. (1967). Hierarchical clustering schemes. *Psychometrika*, **32**, 241–254.

Jolicoeur, P. and Mosimann, J.E. (1960). Size and shape variation in the painted turtle, a principal component analysis. *Growth*, **24**, 339–354.

Jolliffe, I.T. (1972). Discarding variables in a principal component analysis. I. Artificial Data. *Appl. Statist.*, **21**, 160–173.

Jolliffe, I.T. (1973). Discarding variables in a principal component analysis. II. Real Data. *Appl. Statist.*, **22**, 21–31.

Jöreskog, K.G. (1963). Statistical estimation in factor anaysis. Uppsala: Almquist and Wiksell, 145 pp.

Jöreskog, K.G. (1970). A general method for analysis of covariance structures. *Biometrika*, **57**, 239–251.

Jöreskog, K.G. (1977). Factor analysis by least-squares and maximum-likelihood methods. In: Mathematical methods for digital computers Vol. III, edited by K. Enslein, A. Ralston and H.S. Wilf, 125–153. New York; London: Wiley.

Jöreskog, K.G., Klovan, J.E. and Reyment, R.A. (1976). Geological factor analysis. Amsterdam: Elsevier, 178 pp.

Kaiser, H.F. (1958). The varimax criterion for analytic rotation in factor analysis. *Psychometrika*, **23**, 187–200.

Kaiser, H.F. (1962). Formulas for component scores. *Psychometrika*, **27**, 83–86.

Kaiser, H.F. and Caffrey, J. (1965). Alpha factor analysis. *Psychometrika*, **30**, 1–14.

Kass, G.V. (1975). Significance testing in automatic interaction detection. *Appl. Statist.*, **24**, 178–189.

Katz, J.O. and Rohlf, F.J. (1973). Function-point cluster analysis. *Syst. Zool.*, **22**, 295–301.

Katz, J.O. and Rohlf, F.J. (1974). Functionplane – a new approach to simple structure rotation. *Psychometrika*, **39**, 37–51.

Kendall, M.G. (1966). Discrimination and classification. In: Multivariate analysis, edited by P.R. Krishnaiah, 165–185. London; New York: Academic Press.

Kendall, M.G. (1970). Rank correlation methods. London: Griffin, 4th edition.

Kendall, M. (1975). Multivariate analysis. London: Griffin, 210 pp.

Kendall, M.G. and Buckland, W.R. (1971). A dictionary of statistical terms. Edinburgh: Oliver and Boyd, 166 pp.

Kendall, M.G. and Stuart, A. (1968). The advanced theory of statistics. Vol. 3, Design and analysis, and time-series. London: Griffin, 557 pp.

Kendall, M.G. and Stuart, A. (1973). The advanced theory of statistics. Vol. 2, Inference and relationship. London: Griffin, 3rd edition, 723 pp.

Kenkel, N.C. and Orloci, L. (1986). Applying metric and nonmetric multidimensional scaling to ecological studies: some new results. *Ecology*, **67**, 919–928.

Kent, M. and Wathern, P. (1980). The vegetation of a Dartmoor catchment. *Vegetatio*, **43**, 163–172.

Kershaw, K.A. and Looney, J.H.H. (1985). Quantitative and dynamic plant ecology. London: Arnold, 3rd edition, 282 pp.

Kim, J.O. and Mueller, C.W. (1978). Factor analysis. Statistical methods and practical issues. (Sage Univ. Paper Series on Quantitative Applications in the Social Sciences, 07–014). Beverley Hills, Calif.: Sage Publications.

Klovan, J.E. and Imbrie, J. (1971). An algorithm and FORTRAN IV program for large-scale Q-mode factor analysis and calculation of factor scores. *Math. Geol.*, **3**, 61–67.

Knight, A. (1978). Common factor analysis: Some recent developments in theory and practice. *The Statistician*, **27**, 27–42.

Knuth, D.E. (1969). The art of computer programming. Vol. 2, Seminumerical algorithms. London, Ontario: Addison-Wesley, 624 pp.

Koch, G.S. and Link, R.F. (1971). Statistical analysis of geological data. Vol. 2. New York; London: Wiley, 438 pp.

Krishnaiah, P.R. and Lee, J.C. (1977). Inference on the eigenvalues of the covariance matrices of real and complex multivariate normal populations. In: Multivariate analysis IV, edited by P.R. Krishnaiah, 95–103. Amsterdam: North Holland.

Krzanowski, W.J. (1971). Comparison of some distance measures applicable to multinomial data, using a rotational fit technique. *Biometrics*, **27**, 1062–1068.

Kuiper, F.K. and Fisher, L. (1975). A Monte Carlo comparison for six clustering procedures. *Biometrics*, **31**, 777–784.

Lambert, J.M. and Dale, M.B. (1964). The use of statistics in phytosociology. *Adv. Ecol. Res.*, **2**, 59–99.

Lambert, J.M. and Williams, W.T. (1966). Multivariate methods in plant ecology: VI Comparison of information analysis and association analysis. *J. Ecol.*, **54**, 635–664.

Lamont, B.B. and Grant, K.J. (1979). A comparison of twenty-one measures of site dissimilarity. In: Multivariate methods in ecological work, edited by L. Orloci, C.R. Rao and W.M. Stiteler, 101–126. Fairland, Maryland: Int. Co-operative Publishing House.

Lance, G.N. and Williams, W.T. (1967a). A general theory of classification sorting strategies. I. Hierarchical systems. *Comput. J.*, **9**, 373–380.

Lance, G.N. and Williams, W.T. (1967b). Mixed-data classificatory programs. I. Agglomerative systems. *Aust. Comput. J.*, **1**, 15–20.

Lance, G.N. and Williams, W.T. (1968). Note on a new information-statistic classificatory program. *Comput. J.*, **11**, 195.

Lawley, D.N. and Maxwell, A.E. (1971). Factor analysis as a statistical method. London: Butterworth's, 2nd edition, 153 pp.

Lawson, C.L. and Hanson, R.J. (1974). Linear least-squares solutions. New Jersey: Prentice-Hall.

Lazarsfeld, P.L. and Henry, N.W. (1968). Latent structure analysis. Boston: Houghton Mifflin Co.

Lebart, L. (1982). Exploratory analysis of large sparse matrices with application to textual data. In: COMPSTAT 1982 Part 1. Proceedings in computational statistics, edited by H. Caussinus *et al.*, 67–76. Vienna: Physica-Verlag.

Lebart, L. and Fénelon, J.P. (1971). Statistique et informatique appliquées. Paris: Dunod.

Lebart, L., Morineau, A. and Tabard, N. (1977). Techniques de la description statistique. Paris: Dunod.

Lebart, L., Morineau, A. and Warwick, K.M. (1984). Multivariate descriptive statistical analysis. New York; Chichester: Wiley, 231 pp.

Lee, K.L. (1979). Multivariate tests for clusters. *J. Am. Statist. Assn.*, **74**, 708–714.

Lefkovitch, L.P. (1976). Hierarchical clustering from principal co-ordinates: an efficient method for small to very large numbers of objects. *Math. Biosci.*, **31**, 157–174.

Legendre, L. and Legendre, P. (1983). Numerical ecology. Amsterdam; Oxford: Elsevier, 419 pp.

Ling, R.F. (1972). On the theory and construction of *k*-clusters. *Computer J.*, **15**, 326–332.

Linn, R.L. (1968). A Monte Carlo approach to the number of factors problem. *Psychometrika*, **33**, 37–71.

Londo, G. (1971). Patroon en proces in duinvallei vegetaties langs een gegraven meer in de Kennemerduinen. (English summary). Thesis, Nijmegen.

McCammon, R.B. (1966). Principal component analysis and its application in large-scale correlation studies. *J. Geol.*, **74**, 721–733.

McCammon, R.B. (1968). Multiple component analysis and its application in classification of environments. *Bull. Am. Ass. Petrol. Geol.*, **52**, 2178–2196.

McDonald, R.P. (1962). A general approach to non-linear factor analysis. *Psychometrika*, **27**, 397–415.

McDonald, R.P. (1967a). Non-linear factor analysis. Psychometric Monogr. No. 15.

McDonald, R.P. (1967b). Numerical methods for polynomial models in non-linear factor analysis. *Psychometrika*, **32**, 77–112.

McDonald, R.P. (1974). The measurement of factor indeterminacy. *Psychometrika*, **39**, 203–222.

McDonald, R.P. (1977). The indeterminacy of components and the definition of common factors. *Br. J. Math. Statist. Psychol.*, **30**, 165–170.

McDonald, R.P. and Burr, E.J. (1967). A comparison of four methods of constructing factor scores. *Psychometrika*, **32**, 381–401.

MacKay, D.M. (1969). Recognition and action. In: Methodologies of pattern recognition, edited by S. Watanabe, 409–416. London: Academic Press.

Macnaughton-Smith, P. (1963). Predictive attribute analysis. *Biometrics*, **19**, 364–366.

Macnaughton-Smith, P. (1965). Some statistical and other numerical techniques for classifying individuals. Home Office Res. Unit Rep., Publ. No. 6. London: HMSO.

Macnaughton-Smith, P., Williams, W.T., Dale, M.P. and Mockett, L.G. (1964). Dissimilarity analysis: a new technique of hierarchical sub-division. *Nature*, **202**, 1034–1035.

McNeil, D.R. (1977). Interactive data analysis. New York; London: Wiley. 186 pp.

Malcolm, M.A. (1971). On accurate floating-point summation. *Numer. Math.*, **14**, 731–736.

Mansfield, E.R., Webster, J.T. and Gunst, R.F. (1977). An analytic variable selection technique for principal component regression. *Appl. Statist.*, **26**, 34–40.

Manson, V. and Imbrie, J. (1964). FORTRAN program for factor and vector analysis of geologic data using an IBM 7090 or 7094 computer system. *Comput. Contrib. Spec. Publ. Geol. Surv., Kans*, **13**, 46 pp.

Mantel, N. and Valand, R.S. (1970). A technique for nonparametric multivariate analysis. *Biometrics*, **26**, 547–558.

Mardia, K.V., Kent, J.T. and Bibby, J.M. (1979). Multivariate analysis. London; New York: Academic Press, 521 pp.

Marriott, F.H.C. (1971). Practical problems in a method of cluster analysis. *Biometrics*, **27**, 501–514.

Marriott, F.H.C. (1974). The interpretation of multiple observations. London; New York: Academic Press, 117 pp.

Marriott, F.H.C. (1975). Separating mixtures of normal distributions. *Biometrics*, **31**, 767–769.

Martin, R.S., Reinsch, C. and Wilkinson, J.H. (1971). Householder's tridiagonalization of a symmetric matrix. In: Handbook for automatic computation Vol. II Linear algebra, edited by J.H. Wilkinson and C. Reinsch, 212–226. Berlin; Heidelberg; New York: Springer-Verlag.

Martini, I.P. and Acton, C.J. (1975). Use of factor analysis in interpreting genetic processes in lacustrine soils of Ontario, Canada. *Can. J. Earth Sci.*, **12**, 1794–1804.

Massart, D.L. and Kaufman, L. (1983). The interpretation of analytical chemical data by the use of cluster analysis. New York; Chichester: Wiley, 237 pp.

Massy, W.F. (1965). Principal components regression in exploratory statistical research. *J. Amer. Statist. Ass.*, **60**, 234–256.

Mather, P.M. (1976). Computational methods of multivariate analysis in physical geography. London: Wiley, 532 pp.

Matula, D.W. (1977). Graph theoretic techniques for cluster analysis algorithms. In: Classification and clustering, edited by J. Van Ryzin, 95–129. New York: Academic Press.

Ménard, M. and Bélanger, J. (1976). Étude des relations entre les caractères du milieu et la production forestiere par l'analyse factorielle des correspondances. Service de la Recherche, Direction General des Forêts, Ministere des Terres et Forets du Quebec, Mem. No. 24, 73 pp.

Milligan, G.W. (1981). A Monte Carlo study of thirty internal criterion measures for cluster analysis. *Psychometrika*, **46**, 187–199.

Minkoff, E.C. (1965). The effects on classification of slight alterations in numerical technique. *Syst. Zool.*, **14**, 196–213.

Moore, A.W. and Russell, J.S. (1967). Comparison of coefficients and grouping procedures in numerical analysis of soil trace element data. *Geoderma*, **1**, 139–158.

Moore, C.S. (1965). Inter-relationships of growth and cropping in apple trees studied by the method of component analysis. *J. Hort. Sci.*, **40**, 133–149.

Morrison, D.F. (1967). Multivariate statistical methods. New York; London: McGraw-Hill, 338 pp.

Moss, D. (1985). An initial classification of 10-km squares in Great Britain from a land characteristic data bank. *Appl. Geog.*, **5**, 131–150.

Mucina, L. and Polacik, S. (1982). Principal components analysis and trend surface analysis of a small-scale pattern in a transition mire. *Vegetatio*, **48**, 165–173.

Muir, J.W. (1962). The general principles of classification with reference to soils. *J. Soil Sci.*, **13**, 22–30.

Mukkattu, M.M. (1974). Classification of natural communities based on covariance and related functions. *Can. J. Bot.*, **52**, 2341–2349.

Mulaik, S.A. (1976). Comments on 'The measurement of factor indeterminacy'. *Psychometrika*, **41**, 249–262.

Muller, M.W. (1975). A comparison of some clustering techniques. CSIR Special Rep. PERS 224. Johannesburg: Nat. Inst. Personnel Res.

Nabholz, J.V. and Richardson, T.H. (1975). Factor analysis: An exploratory technique applied to mineral cycling. In: Mineral cycling in southeastern ecosystems, edited by F.G. Howell, J.B. Gentry and M.H. Smith, 126–141. U.S. Energy Res. Dev. Admin.

Nichols, S. (1977). On the interpretation of principal components analysis in ecological contexts. *Vegetatio*, **34**, 191–197.

Nishisato, S. (1980). Analysis of categorical data: Dual scaling and its applications. Toronto: Univ. of Toronto Press, 276 pp.

Norris, J.M. (1971). Functional relationships in the interpretation of principal component analysis. *Area*, **3**, 217–220.

Norris, J.M. and Barkham, J.P. (1970). A comparison of some Cotswold beechwoods using multiple discriminant analysis. *J. Ecol.*, **58**, 603–619.

Noy-Meir, I. (1970). Component analysis of semi-arid vegetation in southeastern Australia. Ph.D. Thesis, Australian National Univ.

Noy-Meir, I. (1973). Data transformations in ecological ordinations I. Some advantages of non-centering. *J. Ecol.*, **61**, 329–341.

Noy-Meir, I. and Austin, M.P. (1970). Principal component ordination and simulated vegetational data. *Ecology*, **51**, 551–552.

Noy-Meir, I., Walker, D. and Williams, W.T. (1975). Data transformations in ecological ordination II. On the meaning of data standardization. *J. Ecol.*, **63**, 779–800.

Oksanen, J. (1983). Ordination of boreal heath-like vegetation with principal component analysis, correspondence analysis and multidimensional scaling. *Vegetatio*, **52**, 181–189.

Ord, J.K., Patil, G.P. and Taillie, C. (eds.) (1979). Statistical distributions in ecological work. Fairland, Maryland: Int. Co-operative Publishing House, 464 pp.

Orloci, L. (1967a). An agglomerative method for classification of plant communities. *J. Ecol.*, **55**, 193–206.

Orloci, L. (1967b). Data centering: a review and evaluation with reference to principal component analysis. *Syst. Zool.*, **16**, 208–212.

Orloci, L. (1968). Definitions of structure in multivariate phytosociological samples. *Vegetatio*, **15**, 281–291.

Orloci, L. (1975). Multivariate analysis in vegetation research. The Hague: Junk, 276 pp.

Orloci, L. (1978). Multivariate analysis in vegetation research. The Hague: Junk, 2nd edition, 451 pp.

Packham, J.R. and Willis, A.J. (1976). Aspects of the ecological amplitude of two woodland herbs, *Oxalis acetosella* L. and *Galeobdolon luteum* Huds. *J. Ecol.*, **64**, 485–510.

Page, W.P. and Fabian, R.G. (1978). Factor analysis; An exploratory methodology and management technique for the economics of air pollution. *J. Environ. Manage.*, **6**, 185–192.

Paige, C.C. (1972). Computational variants of the Lanczos method for the eigenproblem. *J. Inst. Math. Applics.*, **10**, 373–387.

Pakarinen, P. and Ruuhijärvi, R. (1978). Ordination of northern Finnish peatland vegetation with factor analysis and reciprocal averaging. *Ann. Bot. Fenn.*, **15**, 147–157.

Patterson, J.G., Goodchild, N.A. and Boyd, W.J.R. (1978). Classifying environments for sampling purposes using a principal component analysis of climatic data. *Agric. Met.*, **19**, 349–362.

Pelto, C.R. (1954). Mapping of multicomponent systems. *J. Geol.*, **62**, 501–511.

Penington, R.H. (1970). Introductory computer methods and numerical analysis. London: Collier-Macmillan, 2nd edition, 497 pp.

Peters, G. and Wilkinson, J.H. (1971). The calculation of specified eigenvectors by inverse iteration. In: Handbook for automatic computation Vol. II, Linear algebra, edited by J.H. Wilkinson and C. Reinsch, 418–439. Berlin; Heidelberg: Springer-Verlag.

Phipps, J.B. (1971). Dendrogram topology. *Syst. Zool.*, **20**, 306–308.

Pielou, E.C. (1969). An introduction to mathematical ecology. New York: Wiley, 266 pp.

Pimentel, R.A. (1979). Morphometrics, the multivariate analysis of biological data. Dubuque, Iowa: Kendall/Hunt, 276 pp.

Pinzka, C. and Saunders, D.R. (1954). Analytic rotation to simple structure. II. Extension to an oblique solution. *Res. Bull.* RB-54-31. Princeton: Educational Testing Service.

Press, S.J. (1972). Applied multivariate analysis. London; New York: Holt, Rinehart and Winston, 521 pp.

Pritchard, N.M. and Anderson, A.J.B. (1970). Observations on the use of cluster analysis in botany with an ecological example. *J. Ecol.*, **59**, 727–747.

Radloff, D.L. and Betters, D.R. (1978). Multivariate analysis of physical site data for wildland classification. *For. Sci.*, **24**, 2–10.

Rao, C.R. (1964). The use and interpretation of principal component analysis in applied research. *Sankhya Ser. A.*, **26**, 329–358.

Rayner, J.H. (1965). Multivariate analysis of montomorillonite. *Clay Minerals*, **6**, 59–70.

Rayner, J.H. (1966). Classification of soils by numerical methods. *J. Soil Sci.*, **17**, 79–92.

Rayner, J.H. (1969). The numerical approach to soil systematics. In: The soil ecosystem (Publication No. 8), edited by J.G. Sheals, 31–39. London: The Systematics Association.

Reinsch, C. and Bauer, F.L. (1971). Rational QR transformation with Newton shift for symmetric tridiagonal matrices. In: Handbook for automatic computation Vol. II, Linear algebra, edited by J.H. Wilkinson and C. Reinsch, 257–265. Berlin; Heidelberg: Springer-Verlag.

Reyment, R.A. (1979). On the interpretation of the smallest principal component. *Bull. Geol. Inst. Univ. Upps. N.S.*, **8**, 1–4.

Reyment, R.A., Berthou, P.Y. and Moberg, B.A. (1976). Statistical recognition of terrestrial and marine sediments in the Lower Cretaceous of Portugal. In: Quantitative techniques for the analysis of sediment, edited by D.F. Merriam, 53–59. Oxford; New York: Pergamon Press.

Rice, J. and Belland, R.J. (1982). A simulation study of moss floras using Jaccard's coefficient of similarity. *J. Biogeogr.*, **9**, 411–419.

Ripley, B.D. (1977). Modelling spatial patterns. *J.R. Statist. Soc. Ser. B.*, **39**, 172–212.

Robertson, P.A. (1979). Comparisons among three hierarchical classification techniques using simulated coenoplanes. *Vegetatio*, **40**, 175–183.

Roger, J.H. (1971). Algorithm AS40. Updating a minimum spanning tree. *Appl. Statist.*, **20**, 204–206.

Roger, J.H. and Carpenter, R.G. (1971). The cumulative construction of minimum spanning trees. *Appl. Statist.*, **20**, 192–194.

Rogers, D.J. and Fleming, H. (1964). A computer program for classifying plants II. A numerical handling of non-numerical data. *Bioscience*, **14**, 15–28.

Rohlf, F.J. (1970). Adaptive hierarchical clustering schemes. *Syst. Zool.*, **19** 58–82.

Rohlf, F.J. (1972). An empirical comparison of three ordination techniques in numerical taxonomy. *Syst. Zool.*, **21**, 271–280.

Rohlf, F.J. (1974a). Graphs implied by the Jardine-Sibson overlapping clustering methods, Bk. *J. Am. Statist. Ass.*, **69**, 705–710.

Rohlf, F.J. (1974b). A new approach to the computation of the Jardine-Sibson Bk clusters. *Comput. J.*, **18**, 164–168.

Rohlf, F.J. (1974c). Methods of comparing classifications. *Annu. Rev. Ecol. Syst.*, **5**, 101–113.

Rohlf, F.J. and Fisher, D.L. (1968). Test for hierarchical structure in random data sets. *Syst. Zool.*, **17**, 407–412.

Romane, F., Guillerm, J.L. and Waksman, G. (1977). Une utilisation possible de l'arbre de portee minimale en phyto-ecologie. *Vegetatio*, **33**, 99–106.

Ross, G.J.S. (1969a). Classification techniques for large sets of data. In: Numerical taxonomy, edited by A.J. Cole, 224–233. London; New York: Academic Press.

Ross, G.J.S. (1969b). Algorithm AS13. Minimum spanning tree. *Appl. Statist.*, **18**, 103–104.

Ross, G.J.S. (1969c). Algorithm AS14. Printing the minimum spanning tree. *Appl. Statist.*, **18**, 105–106.

Ross, G.J.S. (1969d). Algorithm AS15. Single linkage cluster analysis. *Appl. Statist.*, **18**, 106–110.

Rosswall, T. and Kvillner, E. (1978). Principal-components and factor analysis for the description of microbial populations. *Adv. Microb. Ecol.*, **2**, 1–8.

Roux, M. and Salanon, R. (1974). Le role des bryophytes et des lichens dans l'analyse multidimensionelle appliquée a divers groupements a *Pinus sylvestris* L. et *Pinus uncinata* Mill. des Pyrenees orientales, des Causses, et des Alpes austro-occidentales. In 'Les problemes modernes de la bryologie', Colloque Lille, Dec. 1972, 213–224. Paris: Soc. Bot. France.

Rubin, J. (1967). Optimal classification into groups: An approach for solving the taxonomy problem. *J. Theor. Biol.*, **15**, 103–144.

Rummell, R.J. (1970). Applied factor analysis. Evanston, Illinois: Northwestern Univ. Press.

Schiffman, S.S., Reynolds, M.L. and Young, F.W. (1982). Introduction to multidimensional scaling: Theory, methods and applications. London; New York: Academic Press, 413 pp.

Schnell, G.D. (1970). A phenetic study of the suborder Lari (Aves) I. Methods and results of principal components analysis. *Syst. Zool.*, **19**, 35–37.

Scott, A.J. and Symons, M.J. (1971). Clustering methods based on likelihood ratio criteria. *Biometrics*, **27**, 387–397.

Seal, H. (1968). Multivariate statistical analysis for biologists. London: Methuen, 209 pp.

Searle, S.R. (1966). Matrix algebra for the biological sciences. New York; London: Wiley, 296 pp.

Serre, F. (1977). A factor analysis of correspondence applied to ring widths. *Tree-Ring Bull.*, **37**, 21–31.

Shepherd, M.J. and Willmott, A.J. (1968). Cluster analysis on the Atlas computer. *Comput. J.*, **11**, 57–62.

Sibson, R. (1969). Information radius. *Z. Wahrsch' theorie und verw Geb.*, **14**, 149.

Sibson, R. (1971). Some observations of a paper by Lance and Williams. *Comput. J.*, **14**, 156–157.

Sibson, R. (1972). Order invariant methods for data analysis. *J.R. Statist. Soc. Ser. B.*, **34**, 311–349 (with discussion).

Sibson, R. (1973). SLINK: An optimally efficient algorithm for the single link cluster method. *Comput. J.*, **16**, 30–34.

Simonds, J.L. (1963). Application of characteristic vector analysis to photographic and optical response data. *J. Opt. Soc. Am.*, **53**, 968–974.

Smartt, P.F.M., Meacock, S.E. and Lambert, J.M. (1974). Investigations into the properties of quantitative vegetational data. I. Pilot study. *J. Ecol.*, **62**, 735–759.

Smartt, P.F.M., Meacock, S.E. and Lambert, J.M. (1976). Investigations into the properties of quantitative vegetational data. II. Further data type comparisons. *J. Ecol.*, **64**, 41–78.

Sneath, P.H.A. (1964). New approaches to bacterial taxonomy: use of computers. *Ann. Rev. Microbiol.*, **18**, 335–346.

Sneath, P.H.A. (1966). A comparison of different clustering methods as applied to randomly spaced points. *Classn. Soc. Bull.*, **1**, 2–18.

Sneath, P.H.A. (1969). Evaluation of clustering methods. In: Numerical taxonomy, edited by A.J. Cole, 257–271. London; New York: Academic Press.

Sneath, P.H.A. (1977a). A method for testing the distinctness of clusters: a test of the disjunction of two clusters in Euclidean space as measured by their overlap. *J. Math. Geol.*, **9**, 123–143.

Sneath, P.H.A. (1977b). A significance test for clusters in UPGMA phenograms obtained from squared Euclidean distances. *Classn. Soc. Bull.*, **4**, 2–14.

Sneath, P.H.A. (1979). BASIC program for a significance test for clusters in UPGMA dendrograms obtained from squared Euclidean distances. *Comput. and Geosci.*, **5**, 127–137.

Sneath, P.H.A. and Sokal, R.R. (1973). Numerical taxonomy, the principles and practice of numerical classification. San Francisco: W.H. Freeman, 573 pp.

Sokal, R.R. (1959). A comparison of five tests for completeness of factor extraction. *Trans. Kansas Acad. Sci.*, **62**, 141–152.

Sokal, R.R. (1974). Classification: Purposes, principles, progress, prospects. *Science*, **185**, 1115–1123.

Sokal, R.R. (1977). Clustering and classification: Background and current directions. In: Classification and clustering, edited by J. Van Ryzin, 1–15. New York: Academic Press.

Sokal, R.R. and Crovello, T.J. (1970). The biological species concept: a critical evaluation. *Am. Nat.*, **104**, 127–153.

Sokal, R.R., Daly, H.V. and Rohlf, F.J. (1961). Factor analytical procedure in a biological model. *Univ. Kansas Sci. Bull.*, **42**, 1099–1121.

Sokal, R.R. and Michener, C.D. (1958). A statistical method for evaluating systematic relationships. *Univ. Kansas Sci. Bull.*, **38**, 1409–1438.

Sokal, R.R. and Rohlf, F.J. (1962). The comparison of dendrograms by objective methods. *Taxon*, **11**, 33–40.

Sokal, R.R. and Rohlf, F.J. (1969). Biometry: the principles and practice of statistics in biological research. San Francisco: W.H. Freeman, 776 pp.

Sokal, R.R. and Sneath, P.H.A. (1963). The principles of numerical taxonomy. San Francisco; London: Freeman, 359 pp.

Sonquist, J.A. and Morgan, J.N. (1963). Problems in the analysis of survey data and a proposal. *J. Am. Statist. Assn.*, **58**, 415–435.

Sörbom, D. and Jöreskog, K.G. (1976). COFAMM: Confirmatory factor analysis with model modification users' guide. Chicago: National Educational Resources Inc.

Sorensen, T. (1948). A method of establishing groups of equal amplitude in plant sociology based on similarity of species content and its application to analyses of the vegetation on Danish commons. *Biol. Skr.*, **5**, 1–34.

Southwood, T.R.E. (1966). Ecological methods. London: Methuen, 391 pp.

Sparks, D.N. (1973). Algorithm AS58 Euclidean cluster analysis. *Appl. Statist.*, **22**, 126–130.

Sparks, D.N. and Todd, A.D. (1973). A comparison of FORTRAN subroutines for calculating latent roots and vectors. *Appl. Statist.*, **22**, 220–225.

Springall, A. (1978). A review of multidimensional scaling. *Bulletin in Applied Statistics*, **5**, 146–192.

Stephens, W.E. (1976). The display of three-factor models. In: Quantitative techniques for the analysis of sediments, edited by D.F. Merriam, 169–172. Oxford; New York: Pergamon Press.

Swain, C.G., Bryndza, H.E. and Swain, M.S. (1979). Hazards in factor analysis. *J. Chem. Inf. Comput. Sci.*, **19**, 19–23.

Swan, J.M.A. (1970). An examination of some ordination problems by use of simulated vegetational data. *Ecology*, **51**, 89–102.

Taylor, C.C. (1977). Principal component and factor analysis. In: The analysis of survey data Vol. 1. Exploring data structures, edited by C.A. O'Muircheartaigh and C. Payne, 89–124. New York; London: Wiley.

Teil, H. (1975). Correspondence factor analysis: An outline of its method. *Math. Geol.*, **1**, 3–30.

Ter Braak, C.J.F. (1983). Principal components biplots and alpha and beta diversity. *Ecology*, **64**, 454–462.

Ter Braak, C.J.F. (1986). Canonical correspondence analysis: a new eigenvector technique for multivariate direct gradient analysis. *Ecology*, **67**, 1167–1179.

Ter Braak, C.J.F. (1987). The analysis of vegetation-environment relationships by canonical correspondence analysis. *Vegetatio*, **69**, 69–77.

Thompson, J.P. and Rees, R.G. (1979). Pattern analysis in epidemiological evaluation of cultivar resistance. *Phytopathology*, **69**, 545–549.

Thurstone, L.L. (1947). Multiple factor analysis. Chicago: Univ. of Chicago Press, 535 pp.

Tipper, J.C. (1979). An ALGOL program for dissimilarity analysis: A divisive-omnithetic clustering technique. *Comput. and Geosci.*, **5**, 1–13.

Torgerson, W.S. (1958). Theory and methods of scaling. London: Wiley, 460 pp.

Tryon, R.C. and Bailey, D.E. (1970). Cluster analysis. New York: McGraw-Hill, 204 pp.

Tucker, L.R. (1971). Relations of factor score estimates to their use. *Psychometrika*, **36**, 427–436.

Tukey, J.W. (1954). Unsolved problems of experimental statistics. *J. Am. Statist. Assn.*, **49**, 706–773.

Tukey, J.W. (1962). The future of data analysis. *Ann. Math. Statist.*, **33**, 1–67.

Tukey, J.W. (1969). Analyzing data: sanctification or detective work? *Am. Psychologist*, **24**, 83–91.

Väätänen, P. (1980). Factor analysis of the impact of the environment on microbial communities in the Tvarminne area, southern coast of Finland. *Appl. Environ. Microbiol.*, **40**, 55–61.

van Groenewoud, H. (1983). Cluster analysis of simulated vegetation data. *Tuexenia*, No. 3, 523–528.

van der Aart, P.J.M. and Smeenk-Enserink, N. (1975). Correlations between distributions of hunting spiders (Lycosidae, Ctenidae) and environmental characteristics in a dune area. *Neth. J. Zool.*, **25**, 1–45.

van der Maarel, E. (1969). On the use of ordination models in phytosociology. *Vegetatio*, **19**, 21–46.

van der Maarel, E. (1979). Multivariate methods in phytosociology, with reference to the Netherlands. In: The study of vegetation, edited by M.J.A. Werger, 162–225. The Hague: Junk.

Waksman, G., Ménard, M. and Bélanger, J. (1975). Analyse factorielle des relations entre le milieu et la production: étude des tremblaies de la section Laurentienne. *Can. J. For. Res.*, **5**, 662–680.

Wallace, C.W. and Boulton, D.M. (1968). An information measure for classification. *Comput. J.*, **11**, 185–194.

Ward, J.H. (1963). Hierarchical grouping to optimize an objective function. *J. Am. Statist. Ass.*, **58**, 236–244.

Watson, L., Williams, W.T. and Lance, G.N. (1967). A mixed-data numerical approach to Angiosperm taxonomy. *Proc. Linn. Soc.*, **178**, 25–36.

Webb, L.J., Tracey, J.G., Williams, W.T. and Lance, G.N. (1967). Studies in the numerical analysis of complex rain-forest communities: 1. A comparison of methods applicable to site/species data. *J. Ecol.*, **55**, 171–191.

Webb, L.J., Tracey, J.G., Williams, W.T. and Lance, G.N. (1971). Prediction of agricultural potential from intact forest vegetation. *J. Appl. Ecol.*, **8**, 99–121.

Webster, R. (1977). Quantitative and numerical methods in soil classification and survey. Oxford: Clarendon Press, 269 pp.

Webster, R. and Burrough, P.A. (1972a). Computer-based soil mapping of small areas from sample data. I. Multivariate classification and ordination. *J. Soil Sci.*, **23**, 210–221.

Webster, R. and Burrough, P.A. (1972b). Computer-based soil mapping of small areas from sample data. II. Classification smoothing. *J. Soil Sci.*, **23**, 222–234.

Werger, M.J.A. (1978). Vegetation structure in the southern Kalahari. *J. Ecol.*, **66**, 933–941.

Werger, M.J.A., Louppen, J.M.W. and Eppink, J.H.M. (1983). Species performances and vegetation boundaries along an environmental gradient. *Vegetatio*, **52**, 141–150.

Werger, M.J.A., Wild, H. and Drummond, B.R. (1978). Vegetation structure and substrate of the northern part of the Great Dyke, Rhodesia: Gradient analysis and dominance diversity relationships. *Vegetatio*, **37**, 151–161.

Whittaker, R.H. (1967). Gradient analysis of vegetation. *Biol. Rev.*, **49**, 207–264.

Whittaker, R.H. (1970). Communities and ecosystems. New York: Macmillan.

Whittaker, R.H. (1972). Evolution and measurement of species diversity. *Taxon*, **21**, 213–251.

Whittaker, R.H. (ed.) (1973). Ordination and classification of communities. The Hague: Junk.

Wilkinson, C. (1970). Adding a point to a principal co-ordinates analysis. *Syst. Zool.*, **19**, 258–263.

Wilkinson, J.H. (1965). The algebraic eigenvalue problem. London: Oxford Univ. Press.

Wilkinson, J.H. (1978). Singular value decomposition – basic aspects. In: Numerical software, needs and availability, edited by D. Jacobs, 109–135. London; New York: Academic Press.

Williams, E.J. (1952). Use of scores for the analysis of association in contingency tables. *Biometrika*, **39**, 274–289.

Williams, W.T. (1971). Principles of clustering. *Ann. Rev. Ecol. Syst.*, **2**, 303–326.

Williams, W.T. (ed.) (1976). Pattern analysis in agricultural science. Amsterdam; Oxford; New York: Elsevier, 331 pp.

Williams, W.T. and Clifford, H.T. (1971). On the comparison of two classifications of the same set of elements. *Taxon*, **20**, 519–522.

Williams, W.T., Clifford, H.T. and Lance, G.N. (1971). Group-size dependence: a rationale for choice between numerical classifications. *Comput. J.*, **14**, 157–162.

Williams, W.T. and Dale, M.B. (1965). Fundamental problems in numerical taxonomy. *Adv. Bot. Res.*, **2**, 35–68.

Williams, W.T., Dale, M.B. and Lance, G.N. (1971). Two outstanding ordination problems. *Aust. J. Bot.*, **19**, 251–258.

Williams, W.T. and Lambert, J.M. (1960). Multivariate methods in plant ecology. II. The use of an electronic digital computer for association analysis. *J. Ecol.*, **48**, 689–710.

Williams, W.T., Lambert, J.M. and Lance, G.N. (1966). Multivariate methods in plant ecology. V. Similarity analyses and information-analysis. *J. Ecol.*, **54**, 427–445.

Williams, W.T. and Lance, G.N. (1977). Hierarchical classificatory methods. In: Mathematical models for digital computers Vol. III, edited by K. Enslein, A. Ralston and H. S. Wilf, 269–295. New York; London: Wiley.

Williams, W.T., Lance, G.N., Dale, M.B. and Clifford, H.T. (1971). Controversy concerning the criteria for taxonomic strategies. *Comput. J.*, **14**, 162–165.

Williams, W.T., Lance, G.N., Webb, L.J., Tracey, J.G. and Dale, M.B. (1969). Studies in the numerical analysis of complex rain-forest communities. III. The analysis of successional data. *J. Ecol.*, **57**, 515–535.

Williams, W.T., Lance, G.N., Webb, L.J. and Tracey, J.G. (1973). Studies in the numerical analysis of complex rain-forest communities VI. Models of the classification of quantitative data. *J. Ecol.*, **61**, 47–70.

Williamson, M.H. (1978). The ordination of incidence data. *J. Ecol.*, **66**, 911–920.

Wilson, M.V. (1981). A statistical test of the accuracy and consistency of ordinations. *Ecology*, **62**, 8–12.

Wirth, M., Estabrook, G.F. and Rogers, D.J. (1966). A graph theory model for systematic biology, with an example for the Oncidiinae (Orchidaceae). *Syst. Zool.*, **15**, 59–69.

Wishart, D. (1969a). Numerical classification method for deriving natural classes. *Nature*, **221**, 97–98.

Wishart, D. (1969b). Mode analysis: A generalization of nearest neighbour which reduces chaining effects. In: Numerical taxonomy, edited by A.J. Cole, 282–311. London; New York: Academic Press.

Wishart, D. (1969c). An algorithm for hierarchical classifications. *Biometrics*, **25**, 165–170.

Wolda, H. (1981). Similarity indices, sample size, and diversity. *Oecologia*, **50**, 296–302.

Wong, M.A. and Lattin, J.M. (1982). A graph decomposition procedure based on the single-linkage clustering method. *Classn. Soc. Bull.*, **5**, 36–43.

Yates, F. (1949). Sampling methods for censuses and surveys. London: Griffin, 318 pp.

Young, F.W., Takane, Y. and de Leeuw, J. (1978). The principal components of mixed measurement level multivariate data: An alternating least-squares method with optimal scaling features. *Psychometrika*, **43**, 279–281.

Youngs, E.A. and Cramer, E.M. (1971). Some results relevant to choice of sum and sum-of-product algorithms. *Technometrics*, **13**, 657–665.

Yule, G.U. and Kendall, M.G. (1973). An introduction to the theory of statistics. London: Griffin, 14th edition, 701 pp.

Zadeh, L.A. (1977). Fuzzy sets and their application to pattern classification and clustering analysis. In: Classification and clustering, edited by J. Van Ryzin, 251–299. New York; London: Academic Press.

Zelnio, R.N. and Simmons, S.A. (1981). Interpretation of research data: other multivariate methods. *Am. J. Hosp. Pharm.*, **38**, 369–376.

INDEX

249